PALAEONTOLOGICAL AS.
FIELD GUIDES TO FOSSILS: Number 9

Fossils of the Rhaetian Penarth Group

Edited by

ANDREW SWIFT
Palaeobiology Group, Department of Geology,
University of Leicester, University Road,
Leicester LE1 7RH

and

DAVID M. MARTILL
Palaeobiology Research Group,
School of Earth, Environmental and Physical Sciences,
University of Portsmouth, Burnaby Building, Burnaby Road,
Portsmouth PO1 3QL

THE PALAEONTOLOGICAL ASSOCIATION
LONDON
1999

ISBN 0 901702 65 X

Series Editor David K. Loydell
School of Earth, Environmental and Physical Sciences,
University of Portsmouth, Burnaby Building, Burnaby Road,
Portsmouth PO1 3QL

LIST OF CONTRIBUTORS

Barker, Michael J. Palaeobiology Research Group, School of Earth, Environmental and Physical Sciences, University of Portsmouth, Burnaby Building, Burnaby Road, Portsmouth PO1 3QL

Boomer, Ian D. Department of Geography, Daysh Building, University of Newcastle, Newcastle-upon-Tyne NE1 7RU

Duffin, Christopher J. 146 Church Hill Road, Sutton, Surrey SM3 8NF

Hodges, Peter. 41 Parkfields, Pen-y-fai, Bridgend CF31 4NQ

Ivimey-Cook, Hugh C. 6 Lodwells Orchard, North Curry, Taunton TA3 6DX

Jarzembowski, Ed. A. Maidstone Museum and Art Gallery, St Faith's Street, Maidstone, Kent ME14 1LH

Macquaker, Joe H. S. Department of Earth Sciences, University of Manchester, Manchester M13 9PL

Martill, David M. Palaeobiology Research Group, School of Earth, Environmental and Physical Sciences, University of Portsmouth, Burnaby Building, Burnaby Road, Portsmouth PO1 3QL

Munt, Martin C. Museum of Isle of Wight Geology, Sandown, Isle of Wight PO36 8AF

Radley, Jonathan D. Warwickshire Museum, Market Place, Warwick CV34 4SA

Storrs, Glenn W. Cincinnati Museum Center, Geier Collections and Research Center, 1720 Gilbert Avenue, Cincinnati, OH 45202–1401, USA

Swift, Andrew. Palaeobiology Group, Department of Geology, University of Leicester, Leicester LE1 7RH

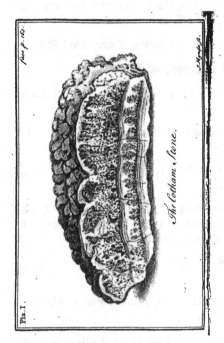

face p. 163.

J. Mynde fc.

The Cotham Stone.

Pla. I.

[163]

OBSERVATIONS

On the EARTHS, ROCKS, STONES,
and MINERALS, about *Briſtol,* &*c.*

●●

B O O K the FIFTH.
Containing an account of certain STONES
of a very particular kind.

●●

CHAPTER the FIRST.
On the Cotham *Stone.*

 HAVE taken the liberty of
making uſe of a new name in the
title of this chapter, but it is upon a very fair occaſion. It has
always been allowed people who were to
ſpeak of things which, not being known
before,

Frontispiece. Illustration of Cotham Stone from E. Owen's 1754 book
*Observations on the Earth, Rocks, Stones and Minerals, for some miles about
Bristol, and on the nature of the Hot-well and the virtues of its waters.* This is
possibly the oldest illustration of a fossil from the Penarth Group.

CONTENTS

ACKNOWLEDGEMENTS

During the writing and compilation of this volume the editors and contributors have profited from many discussions on the geology and palaeontology of the Penarth Group. In particular we must record our thanks to the many reviewers and referees who helped us to improve this volume. The impetus for this guide came from the enthusiasm of the members of the Palaeontological Research Group led by the editors for the Department of Adult Education at Leicester University during the winter of 1994/95, and we thank them for their data-gathering efforts at the commencement of this work. We also thank David Loydell for his overall editing on behalf of the Palaeontological Association.

We are indebted to several institutions and their staff for allowing access to collections and for generous loans of specimens and photographs, in particular: Paul Elkin and R. C. L. Draper at the Bath Royal Literary and Scientific Institute; Sandra Chapman, John Cooper, P. Crabb, Jill Darrell, Tiffany Foster, David Lewis, Sarah Long, Giles Miller, Andrew Ross, Andrew Smith, Stuart Sutherland, Sally Young and library staff at The Natural History Museum, London; Steve Tunnicliff at the British Geological Survey, Keyworth, Nottingham; Mike Dorling at the Sedgwick Museum, Cambridge; Tom Sharpe at the Department of Geology at the National Museum of Wales, Cardiff; Liz Loeffler at the Department of Geology, University of Bristol; Roger Clark at the Bristol City Museum and Art Gallery; John Martin at the New Walk Museum, Leicester; Roy Clements at the Department of Geology, University of Leicester and R. Stutely from the Maidstone Museum and Art Gallery. We also offer our grateful thanks to the above institutions for permission to photograph and figure specimens in their care.

Andrew Swift is grateful to Hugh Ivimey-Cook for making available his wide experience of the Penarth Group, to Terry Baldwin, Noel Worley and British Gypsum Ltd and to Rugby Cement for permission to visit and figure specimens from Bantycock Quarry, Newark and Southam Works (Long Itchington) Quarry respectively, and to Joanne E. Norris for assistance and support at various times during the preparation of this field guide. Dave Martill thanks P. Blake, M. Curtis, A. Hunt, G. Long, R. Loveridge, D. Pemberton and A. Walsh for assistance with his contributions. Ed. A. Jarzembowski acknowledges Royal Society grant no. 18525.

1. INTRODUCTION – THE PENARTH GROUP

by ANDREW SWIFT and DAVID M. MARTILL

The Penarth Group is a thin but diverse suite of sedimentary rocks with a narrow, south-west–north-east trending outcrop from the south Devon and Dorset coast to the mouth of the Tees in north-east England (Text-fig. 1). Minor outliers occur west of the main outcrop in the Solway Basin, near Carlisle (Ivimey-Cook et al. 1995), the Cheshire Basin (Poole and Whiteman 1966), the Needwood Basin in north Staffordshire (Stevenson and Mitchell 1955) and at Copt Heath near Birmingham (Eastwood et al. 1925) (Text-fig. 1). Sedimentary rocks attributed to, or considered to be equivalent to, the Penarth Group also occur in scattered outcrops in the Scottish Western Isles and on Arran (Judd 1878; Peach et al. 1901; Lee 1920; Bailey and Anderson 1925; Lee and Bailey 1925; Tyrrell 1928; Lee and Pringle 1932; Tomkieff 1969), as well as in Northern Ireland (Tate 1864, 1867a; Lamplugh et al. 1904; Fowler and Robbie 1961; Manning et al. 1970; Ivimey-Cook 1975; Manning and Wilson 1975; Wilson and Manning 1978; Griffith and Wilson 1982). Certain sections on the eastern coast of Scotland, e.g. the large exotic block at Elgin (Taylor and Cruickshank 1993), have also been suggested as having a Rhaetian age, but the detailed stratigraphy of these isolated sequences has yet to be securely established.

Despite the limited area of outcrop, rocks of the Penarth Group or coeval strata occur extensively beneath the surface on- and offshore of the British Isles. In boreholes east of the main outcrop the Penarth Group has been detected at depth extending to the presumed margins of the basin in south-east England (Lees and Cox 1937; Lees and Taitt 1946; Falcon and Kent 1960; Gallois 1965; Gallois and Worssam 1993), and coeval marginal deposits are recorded lapping on to the edges of the London-Brabant land mass (Donovan et al. 1979). The group and its equivalents have also been proved beneath east and north-east England south of the Tees (e.g. Gaunt et al. 1980) and the subcrop extends beneath the North Sea as proved by hydrocarbon investigations (e.g. Geiger and Hopping 1968). The Penarth Group and contemporary deposits also occur extensively off the coasts of south and south-west Britain (e.g. Hamblin et al. 1992; Tappin et al. 1994).

Historically the Penarth Group sequence was referred to as the 'Rhaetic', after the type area of the Tethyan Triassic deposits in the Rhaetic Alps of Austria with which the British rocks were compared. The deposits are still often referred to as the 'Rhaetic', but this is a practice that should now be abandoned, as confusion has been generated by the

TEXT-FIG. 1. Outcrop of the Penarth Group in England and Wales, showing chief localities cited in the text.

indiscriminate use of the name in both chronostratigraphical and litho-stratigraphical contexts. The Penarth Group is a strictly lithostratigraphical term, and avoids these dual nomenclatural problems (Warrington *et al.* 1980). There has been a long-standing debate regarding the precise age of the Penarth Group, and whether it represents the first unit of strata within the Jurassic or the last within the Triassic, or perhaps even straddles the boundary. Resolution of the debate has been frustrated by the distinctive nature of the sediments and the absence of diagnostic zone fossils. Recent palynological and conodont studies, however, have established that the group lies wholly within the Rhaetian Stage of the latest Triassic (Warrington *et al.* 1980; Swift 1989, 1995a, 1995b; but see Donovan *et al.* 1989), and present estimates indicate that it was deposited between 210–208 Ma (Harland *et al.* 1990).

The best exposures of the Penarth Group occur in coastal sections, in unstable sea cliffs on the Dorset/Devon coast, in north Somerset and South Wales (Pls 1–5). In all these areas, coastal and climatic erosion are constantly changing the configuration of the exposures and rock falls and slumps frequently obscure the sequence to a greater or lesser extent. The group is also well shown at several picturesque river cliffs on the banks of the River Severn (Pl. 6). It is these natural coastal and river cliff localities which have yielded most of our knowledge of the nature of the Penarth Group and its fossils, but man-made inland sections exposed in quarries and road and rail cuttings have also produced important fossil finds and stratigraphical information. Natural exposures of the Penarth Group are rare inland, as the group is dominated by easily weathered soft clastic lithologies, but thin limestones and harder beds developed at some levels do give rise to small topographic features, including low escarpments and benches, notably in the south-central Midlands, north Leicestershire and south Nottinghamshire. From north Nottinghamshire to the point where the Penarth Group meets the sea near the mouth of the River Tees the outcrop is masked by thick glacial and superficial deposits.

Rocks around Bath now attributed to the Penarth Group were mapped in manuscript form as early as 1799 by the founding father of British stratigraphy, William Smith. Since that time there have been numerous descriptions of the stratigraphy and accounts of the palaeontology of the Penarth Group as the 'Rhaetic', but no complete synthesis has been presented. The latter half of the nineteenth and early years of the twentieth century saw the publication of several key studies on the Penarth Group, by such locally eminent geologists as Thomas Wright (e.g. 1860), Charles Moore (e.g. 1861*b*), Peter Bellinger Brodie (e.g. 1876, 1886) and, notably, Linsdall Richardson with more than 25 publications relating to the Penarth Group between 1901 and 1930 (e.g. major publications in 1905 and 1911). Moore's collection, which formed the basis of the first serious systematic study of Penarth Group fossils (Moore 1861*b*), is of national importance and is housed in the museum of the Bath Royal Literary and Scientific Institute. Other collections are dispersed between several institutions, with some of the most extensive in the British Geological Survey collections and in The Natural History Museum, London.

The deposits of the Penarth Group record a complex pattern of sedimentation which evolved over a relatively short period due to the effects of several minor transgressions and regressions which occurred within the major Late Triassic world-wide marine trangression. These strata contain the fossil remains of a fascinating assemblage of animals, both marine and terrestrial, vertebrate and invertebrate, which inhabited the specialized environments thus created, including very rare, tiny teeth of some of the earliest mammals. The group has also attained prominence palaeontologically for the widespread occurrence at the base of the succession of bone beds rich in fish and reptile remains (see chapters 4, 5, 14, 15). Furthermore, the Penarth Group is important in preserving the first British representatives of faunas which herald the great Mesozoic Marine Revolution. These fossils constitute the first truly marine assemblages in Britain since the Late Permian, some forty-five million years earlier, and occur at a critical time of biotic turnover associated with the end-Triassic extinction (Benton 1990).

Palaeokarst topography was developed on exposed Carboniferous Limestone during deposition of the Penarth Group in the Mendips and South Wales, and this landscape contained fissures and cave systems which became filled with a confusing mélange of sediments, some of which are in part coeval with the Penarth Group and in close proximity to

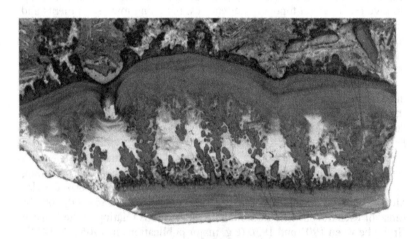

TEXT-FIG. 2. Polished slab of Cotham or Landscape Marble. From Cotham, Bristol; ×1. See also Chapter 17, Plate 32, figure 9.

it (Savage 1993). These deposits include bone breccias which are of palaeontological importance for the abundance and diversity of their faunas, including lizards, rhynchocephalians, dinosaurs, pterosaurs and the teeth of small mammals. However, due to the lack of stratigraphical resolution of these atypical deposits, their biota has been excluded from this field guide except where the host horizon is reasonably stratigraphically secure or where a taxon is known to occur also within the main body of the Penarth Group.

Currently the Penarth Group is of little economic importance, but was previously valuable locally. Its clays have been used in brick manufacture and its limestones have been and still are used for cement production. Other applications for the hard micritic limestones of the Langport Member ('White Lias') have included use as a building stone and for road aggregate. A small cottage industry producing ornamental pictures from an algal stromatolite known as the Cotham or Landscape Marble developed for a short period in the last century. This took the form of slicing thin blocks of the limestone to reveal an arborescent pattern resembling a landscape (Text-fig. 2; also Pl. 32, fig. 9). To this was usually added a small sketch of a village.

This guide is intended to give as comprehensive a coverage as possible of the animal and plant fossils known to date from the Penarth Group, with the exception of palynomorphs. Descriptions and discussions form the bulk of this book and will be found in Chapters 6–17, which cover the various fossil groups. Chapters 1–3 comprise an introduction to the Penarth Group followed by summaries of the stratigraphy and history of palaeontological studies, whilst chapters 4 and 5 look specifically at the genesis of the bone beds and the sedimentology of the remarkable series of beds which constitute the Westbury Formation.

2. STRATIGRAPHY (INCLUDING BIOSTRATIGRAPHY)

by ANDREW SWIFT

The Penarth Group is divided into two distinct formations on lithological grounds: a lower Westbury Formation, overlain by the Lilstock Formation; both units can be traced throughout the outcrop (Text-fig. 3). The Lilstock Formation is divided into two members, a lower Cotham Member and an upper Langport Member (Warrington *et al.* 1980). Rocks of the Westbury Formation and the Cotham Member are widespread, but Langport Member strata are most characteristically developed in south-west and south central England, with only thin, impersistent representatives of the member extending into Nottinghamshire and beyond into northern England (Swift 1995*a*).

Chronostratigraphic Units		Lithostratigraphic Units			
Jurassic	Hettangian	Lias Group	Beds with *Psiloceras*		
			Pre-planorbis Beds		
			? 'Watchet Beds' (north Somerset/South Wales only)		
Triassic	? Rhaetian	Penarth Group	Lilstock Formation	Langport Member	
				Cotham Member	
			Westbury Formation		
		Mercia Mudstone Group	Blue Anchor Formation	'Grey Marls' or 'Sully Beds' (north Somerset/South Wales only)	

TEXT-FIG. 3. British Late Triassic–Early Jurassic stratigraphy (after Warrington *et al.* 1980 and Swift 1995*b*).

PLATE 1

1

2

3

In continuous successions the Penarth Group is usually between 5 m and 20 m thick and typically rests disconformably on argillaceous beds of the Blue Anchor Formation of the Mercia Mudstone Group. In north Somerset and South Wales the higher beds of the Blue Anchor Formation record marine influence and have been given local names (Mayall 1981). In some places in south-west England and South Wales the Penarth Group overlies Carboniferous and older rocks, often with angular unconformity (Callaway 1901). The upper boundary of the Penarth Group is usually marked by a disconformity with the overlying Preplanorbis Beds of the Lias Group or later beds, but in sections on the north Somerset coast and in South Wales, an uninterrupted passage through the shales of the 'Watchet Beds' (Watchet Member of Hallam 1990) into the Preplanorbis Beds has been postulated (Text-fig. 3). Commonly, a pause in deposition or an erosion surface can be detected at the upper boundary, e.g. in Long Itchington (Southam) Quarry near Rugby [SP 420 632] (Text-fig. 1; Pl. 1, figs 1–2), where clays, probably of the *Schlotheimia angulata* Zone (Early Jurassic), rest directly on a very irregular hardground surface of Langport Member limestone, with the Preplanorbis Beds and the two lowest Lias Group ammonite zones missing. A full succession is well exposed at a number of localities, but perhaps most impressively in the sea cliffs of the group's type area between Penarth [ST 192 719] (Text-fig. 1; Pl. 1, fig. 3) and St Mary's Well Bay [ST 178 677] (Text-fig. 1; Pl. 2, figs 1–2), near Cardiff in South Wales.

In Scotland the Penarth Group is represented only with certainty by a few scattered short sequences in the Western Isles and on Arran, where field relations are often hard to unravel. Where boundaries can be seen, the group is underlain by the Mercia Mudstone Group and overlain discomformably by the Lias Group or its equivalents. However, on Arran, a shattered Penarth Group succession was caught up and incorporated into a Tertiary igneous suite and is partially preserved as large isolated fragments on the flanks of the Central Ring Complex (Peach *et al.* 1901). In Northern Ireland the group and its equivalents are similarly exposed only in short incomplete sequences around the margins of the Antrim Plateau, overlain by either the Lias Group, Cretaceous sedimentary rocks

---

EXPLANATION OF PLATE 1

Fig. 1. General view of Long Itchington (Southam Works) Quarry, near Rugby, Warwickshire. The Langport Member forms the prominent bench in the foreground.

Fig. 2. The Langport Member succession at Long Itchington (Southam Works) Quarry, near Rugby, Warwickshire.

Fig. 3. General view of Penarth Bay, South Glamorgan, with the town of Penarth in the right background.

PLATE 2

or by Paleocene plateau basalts (Manning *et al.* 1970). Despite the geographical separation of the Scottish, Irish and mainland Penarth Group outcrops, sediments at some horizons are remarkably similar in all areas, particularly those of the Westbury Formation.

The three units of the Penarth Group, i.e. the Westbury Formation and Cotham and Langport members of the Lilstock Formation, have a composite type section drawn from three coastal exposures situated to the west of the town of Penarth in South Wales (Warrington *et al.* 1980). These are at Penarth Head [ST 192 719], near Lavernock Point [ST 188 682] (Text-fig. 1; Pl. 2, figs 3–4) and at St Mary's Well Bay [ST 178 677] (Text-fig. 1; Pl. 2, figs 1–2).

WESTBURY FORMATION

The lowest unit of the Penarth Group, the Westbury Formation, takes its name from a locality at Garden Cliff, near Westbury-on-Severn [SO 718 127] (Text-fig. 1; Pl. 6, fig. 1). In most areas the Westbury Formation constitutes the first expression of the Late Triassic marine transgression in Britain, but on the north Somerset coast and in South Wales the higher beds (variously termed 'Grey Marls', 'Sully Beds' or Williton Member of Mayall 1981) of the underlying Blue Anchor Formation of the Mercia Mudstone Group demonstrate an earlier marine influence by the occurrence of marine bivalves and other taxa. This is consistent with a transgression from the south moving gradually northwards. The junction between the two units is usually disconformable and it is not unusual to see clasts of Blue Anchor lithologies incorporated in basal Westbury Formation beds. The ultimate expression of this is the spectacular basal bone bed of Aust Cliff, Gloucestershire [SO 565 896] (Text-fig. 1; Pl. 6, fig. 2), which incorporates clasts of Blue Anchor Formation lithologies up to 500 mm long (see Chapter 5).

The consistency of the character of the formation throughout most of its outcrop is remarkable and argues for a generally flat peneplained topography prior to inundation, which was rapidly flooded. A fairly consistent thickness of 4–10 m is present, except where the unit approaches the

---

EXPLANATION OF PLATE 2

Fig. 1. Detail of Westbury Formation–Langport Member succession at St Mary's Well Bay, South Glamorgan.

Fig. 2. Detail of Cotham Member–lower Lias Group succession at St Mary's Well Bay.

Fig. 3. The cliff at Lavernock Point, South Glamorgan, showing the Penarth Group succession.

Fig. 4. Detail of Cotham Member–'Watchet Beds' succession at Lavernock Point.

margins of the basin and onlaps against positive areas such as the London Platform or the Carboniferous Limestone 'islands' in the Mendips. In these cases contemporary sediments consist of thinner, coarser deposits, but over the main outcrop the Westbury Formation is characterized by thin beds of soft, easily weathered shales, some of which are 'papery', and mudstones, with subsidiary thin silt- and sandstones. Some laterally impersistent limestones occur in the more southerly parts of the outcrop, but become progressively rarer northwards. The most striking feature of the argillaceous beds is their dark grey, almost black, coloration. Enhanced, but not excessively high, organic carbon levels are largely responsible (Tuweni and Tyson 1994; but see Macquaker 1994), but there is also a component of finely disseminated pyrite. The shallow water origin of the sediments is demonstrated by sedimentary structures such as ripple marks, which may be developed in the sandier layers, and an impressive suite of trace fossils characteristic of shelf environments (see Chapter 16). The well-known bone beds (see Chapter 5), also support a shallow water origin for the sediments, the finer muds being deposited more distally from sediment sources.

The unusual sediments and the fossils of the Westbury Formation record distinctive environments. The waters which flooded the basin were marine, but were subject to fluctuating salinity during deposition of the formation. This resulted from their being contained within an isolated basin with poor or intermittent connections with the open sea, restricted circulation and probably poorly oxygenated bottom waters. These fluctuations give a roughly cyclic character to the beds (Duffin 1980; Macquaker 1994, this volume). The invertebrate animals which were able to colonize this relatively hostile environment were low in diversity and rather small in size. Some groups are not represented at all, such as sponges and bryozoans; only one specimen of a cephalopod has been found, a small ammonite (Donovan *et al.* 1989), its provenance remaining unsupported by other finds. Only two inarticulate brachiopod genera have been recorded, one rare and the other extremely rare. Often large numbers of a single species of bivalve are found crowded on a bedding plane, whilst surrounding levels are barren, suggesting the rapid termination of conditions suitable for the organisms and the establishment of a harsh

---

EXPLANATION OF PLATE 3

Fig. 1. The cliff sequence from 'Grey Marls' (bottom left)–lower Lias Group *c*. 1 km east of the beach access at Lilstock, north Somerset.

Fig. 2. The complete exposed sequence at Bantycock Quarry, near Newark, Nottinghamshire. The dark beds around the centre of the sequence form the Westbury Formation, with the slabby limestones of the Lias Group at the top of the cliff.

PLATE 3

1

2

environment. This can be seen throughout the Westbury Formation and is clearly related to the cyclicity evident in the beds.

LILSTOCK FORMATION

The type area for the Lilstock Formation is the north Somerset coast and South Glamorgan, with representative sections at Lilstock [ST 177 454] (Text-fig. 1; Pl. 3, fig. 1) and Lavernock Point [ST 187 682] (Warrington *et al.* 1980) (Text-fig. 1; Pl. 2, figs 3–4). Thickness is very variable over the outcrop, particularly in the higher unit, the Langport Member, and the formation in the type area is atypically thin e.g. at Lilstock it is only *c.* 3 m, whereas in contrast, at Bantycock Quarry in Nottinghamshire [SK 813 495] (Text-fig. 1; Pl. 3, fig. 2), the Cotham Member alone is almost 7 m thick. At Pinhay Bay near Lyme Regis [ST 319 909] (Text-fig. 1; Pl. 4, figs 1–2) the Langport Member is more than 7 m thick and in a borehole at Westbury, Wiltshire (Pringle 1922), it proved to be 8 m thick. Further north however, the Langport Member disappears or attenuates markedly (Swift 1995a), even in areas of thick Cotham Member deposition.

Cotham Member
The Westbury Formation is succeeded by the lowest unit of the Lilstock Formation, the Cotham Member, which takes its name from the village of Cotham [ST 58 74], now a suburb of Bristol (Text-fig. 1). It is not possible everywhere to determine the exact relationship with the underlying Westbury Formation or the point of demarcation, but, despite the conformable nature of the junction, a change in depositional mode can usually be discerned by a sudden colour change or an irregular contact. However, in northern parts of the outcrop this colour change may be gradual, demonstrating some continuity of deposition. Thickness is variable, rarely exceeding 8 m, but in contrast with the succeeding Langport Member, an apparent thickening northwards is discernible. The reasons for this are

---

EXPLANATION OF PLATE 4

Fig. 1. The Langport Member–lower Lias Group sequence at Pinhay Bay, near Lyme Regis, showing the strong tonal contrast between the two units.

Fig. 2. Detail of the Langport Member sequence at Pinhay Bay showing soft sediment deformation features.

Fig. 3. General view of the small bay at St Audrie's, east of Watchet, north Somerset.

Fig. 4. Foreshore reefs at St Audrie's Bay. The hammer rests on the bottom bed of the Langport Member.

Fig. 5. A deformed/slumped bed in the Cotham Member at St Audrie's Bay.

PLATE 4

unclear, but it may be the result of reduced clastic input and slower sedimentation rates in more basinward southern areas. As with the Westbury Formation, a cyclicity to the patterns of sedimentation in the Cotham Member has been recognized (Duffin 1980).

Like the Westbury Formation, the Cotham Member consists mainly of fine-grained clastic sedimentary rocks such as shales and fine sandstones, but they are slightly coarser-grained overall and lighter in colour, being pale greenish-grey with bluish tones at certain levels. In the Midlands and Lincolnshire a reddish brown colour is sometimes apparent in the rocks, suggesting the re-establishment of arid, oxidizing conditions similar to those apparent during deposition of the 'Keuper Marls' of the underlying Mercia Mudstone Group. In contrast with the Westbury Formation a significant calcareous component is often present in Cotham Member strata, indicating greater chemical control to sedimentation. Limestones and hard calcareous sandstones are common, especially in the southerly parts of the outcrop. Sedimentary features such as lenticular bedding, ripple-marks and cross-bedding are ubiquitous and indicate very shallow water and dominantly moderate-high energy conditions. Hardground surfaces occur and occasional emergence is demonstrated by desiccation cracks, which are impressively exposed at Lavernock Point, where they extend down into the underlying Westbury Formation. Seismic activity most probably associated with the opening of the proto-Atlantic Ocean and the re-activation of faults led to the widespread formation of slumped deposits on both a large and small scale. This is particularly well seen in the Cotham Member exposed in St Audrie's Bay [ST 103 432] (Text-fig. 1; Pl. 4, figs 3–5), and is also observable in the Penarth Group outcrops in Northern Ireland.

Conditions changed markedly during deposition of the Cotham Member and it is apparent that the basin became largely isolated from the marine source which had fed the Westbury Formation sea. As a consequence of this regression the biota was adversely affected and fossils are scarce in the Cotham Member. But, the dynamic, rapidly changing nature of the conditions controlling deposition of the Penarth Group meant that even within this largely non-marine regime, periodic short-duration marine pulses allowed a partial recolonization of the basin, demonstrated by the occurrence at certain levels of marine bivalves and microplankton. This phenomenon, however, is mainly restricted to the southerly portion of the outcrop, which lay closer to an oceanic source of marine waters. Contemporary rocks in Northern Ireland commonly contain a range of marine bivalves at levels throughout the Cotham Member (Ivimey-Cook 1975), indicating more continuous marine influence. Towards or at the top of the Cotham Member in southern England occurs the famous algal deposit known as the Cotham or Landscape Marble (Text-fig. 2; Pl. 32, fig. 9; front cover).

Langport Member

The onset of another marine transgression and renewed connections with the main Tethyan ocean lying to the south of the present British Isles terminated Cotham Member deposition and initiated deposition of the Langport Member of the Lilstock Formation. The member takes its name from the town of Langport in Somerset [ST 410 264–430 277], around which the member was previously quarried extensively. The junction of the Cotham and Langport members is conformable, but usually sharp, indicating a period of erosion prior to or perhaps after the onset of Langport Member sedimentation.

The character of the deposits of the Langport Member is in marked contrast with the other main transgressive sediments of the Penarth Group represented by the Westbury Formation, with its dark, non-calcareous, clastic strata. The rocks of the Langport Member are characteristically light-coloured, probably clear-water, chemically precipitated micrites with a low clastic component, and subsidiary calcareous shales. It is clear that environmental conditions were markedly different from those pertaining at the time of the deposition of the Westbury Formation. Clastic input from surrounding land areas had been reduced, suggesting a drier climate with less runoff and a much-reduced drainage system feeding into the basin. This climatic regime probably began earlier, during deposition of the Cotham Member, and led to changes in the biota. Corals, extremely rare in the Westbury Formation, are far more widely distributed and numerous in the Langport Member, where the clearer water conditions were evidently more suitable. Other groups also occur for the first time in the Penarth Group, such as conodonts, which may have favoured warm, well-oxygenated waters. Despite these disparities with the Westbury Formation, there is evidence to suggest that conditions in the waters in which the sediments of the Langport Member were laid down were also not ideal for the proliferation of a varied and abundant fauna. This is reflected in the rather restricted number of taxa present and their patchy occurrence, with several groups being represented by a handful of species and others not present at all, e.g. there are no brachiopods, cephalopods or sponges, and no arthropods with the exception of ostracods. It is probable that the Langport Member 'sea' had only restricted or intermittent connections with the main ocean, and this may in turn have led to abnormal and varying salinities. Very shallow water and frequent periods of exposure would also prove to be inhibiting factors to colonization. The most typical fossils are dominantly shallow-water, attached bivalves such as oysters and mussels.

That the sea was very shallow at the time of deposition of the Langport Member is shown by an abundance of features such as desiccation cracks, rill marks and encrusting bivalves. Many of the limestone beds have very uneven, eroded and reddened upper surfaces, indicating regular periods of

PLATE 5

1

2

3

4

emergence and exposure. The whole unit is characterized by these discontinuity surfaces. The top bed has been named the 'sun bed' in the south-west, reflecting its dried and cracked appearance. Pervasive bioturbation is evident in many sequences, particularly in the south central Midlands. In Vallis Vale [ST 761 497], near Frome, (Text-fig. 1; Pl. 5, fig. 1) the uppermost bed of the Langport Member is riddled with the borings of *Lithophaga* bivalves (Pl. 16, fig. 6), which were almost certainly excavated in the Bajocian (Mid Jurassic), reflecting the considerable unconformity and onlap close to the Mendip Carboniferous Limestone 'islands'. The ubiquitous occurrence of fine, comminuted, indeterminate plant material throughout the sequence demonstrates that at no point was the area of deposition far from land.

The thickness of the Langport Member is extremely variable, even over short distances, from 0–8 m, but such variations are consistent with the evidence of shallowness, fluctuating sea level and periods of sub-aerial exposure. The 'islands' and 'highs' which were effective in Rhaetian times in south-west England and South Wales also exerted an influence on sedimentation. Under such a regime, periods of non-deposition and erosion occurred on both a large and small scale. Average water depth was probably no greater than 10 m anywhere.

The Langport Member, unlike the underlying Westbury Formation and Cotham Member, shows a general reduction in thickness northwards and becomes intermittent in its occurrence before petering out completely in its typical 'White Lias' facies in north Nottinghamshire/Lincolnshire (Swift 1995a). It has been suggested that this facies may be replaced in northerly areas by contemporary thin limestones and shales of Liassic aspect (Warrington and Ivimey-Cook 1992).

The soft sediment deformation first apparent in the Cotham Member persists into the Langport Member, sometimes to impressive effect. This is best seen on both a large and small scale in the slumped and re-deposited

---

EXPLANATION OF PLATE 5

Fig. 1. The basal beds of the attenuated, conglomeratic Penarth Group sequence at Vallis Vale, near Frome, Somerset. Below the cleft (but unseen in the photograph) towards the bottom of the picture are peneplained beds of Carboniferous Limestone.

Fig. 2. The main cliff exposure of Langport Member at Culverhole Point, east of Seaton, Devon.

Fig. 3. A slipped block on the beach at Culverhole Point showing the relatively undeformed basal beds of the Langport Member overlain by heavily contorted and brecciated strata.

Fig. 4. The Langport Member sequence at Dark Lane Copse Quarry, Lighthorne, Warwickshire. The hammer rests on the basal bed.

1

2

beds at Pinhay Bay (Hallam 1960) (Text-fig. 1; Pl. 4, figs 1–2) and Culverhole Point [SY 273 894] (Text-fig. 1; Pl. 5, figs 2–3), but disturbed and brecciated beds also occur in the Midlands e.g. at Dark Lane Copse Quarry, Warwickshire [SP 325 558] (Pl. 5, fig. 4).

In north Somerset and South Wales a sequence of generally dark, blocky or fissile shales and fine siltstones intervenes between the top of the 'White Lias' limestones of the Langport Member and the 'paper shales' of the basal Lias Group. These deposits, which are normally informally called the 'Watchet Beds', have remained stratigraphically 'in limbo', although a recommendation of the Geological Society Working Group (Warrington *et al.* 1980) is to subsume them into the Langport Member, thereby designating them, where developed, as the highest beds in the Penarth Group. However, their lithology is in marked contrast with the underlying micritic limestones and pale calcareous shales, and an equally strong case would seem to be in place for allocation to the Lias Group, or indeed for separation as a distinct member, as suggested by Hallam (1990).

BIOSTRATIGRAPHY

Fossils helpful to the elucidation of the stratigraphy of the Penarth Group are very few, the almost total lack of ammonites (one specimen only, see Donovan *et al.* 1989 and Chapter 12) being the chief barrier to accurate correlation with major sequences elsewhere. However, a number of other fossil groups has been utilized with varying degrees of success. Wright (1860) was the first to attempt to define a biostratigraphy for the sequence of beds now attributed to the Penarth Group. He erected the 'Zone of *Avicula* (=*Rhaetavicula*) *contorta*', a definition which originally included only those strata now attributed to the Westbury Formation and Cotham Member of the Lilstock Formation. This zone was later expanded by Moore (1861*b*) to include also those beds which now comprise the Langport Member; this definition effectively united for the first time the beds which now comprise the Penarth Group (Text-fig. 3). *Rhaetavicula contorta* remains the only macrofossil with proven international correlative application, occurring in both Tethyan and Penarth Group sequences.

EXPLANATION OF PLATE 6

Fig. 1. View along Garden Cliff, Westbury-on-Severn, Gloucestershire. Nearest the camera is the Blue Anchor Formation-Westbury Formation succession.

Fig. 2. Mercia Mudstone Group–lower Lias Group succession at Aust Cliff, Gloucestershire, showing prominent fault. The lowest two-thirds of the cliff consists of red, gypsiferous 'Keuper Marl' of the Mercia Mudstone Group.

Attempts at establishing a zonation using potentially more useful microfossils came much later when Anderson (1964) erected a series of four zones based on ostracod assemblages. This scheme has rarely been utilized but was updated with systematic revisions by Bate (1978*a*). In ascending order the ostracod zones are characterized by *Rhombocythere penarthensis* in the 'Lower Rhaetian', *Ogmoconcha bristolensis* and *O. martini* defining the 'Middle Rhaetian' and *O. moorei* defining the 'Upper Rhaetian' (Bate 1978*a*). Later, Lord and Boomer (1990) reported that at St Audrie's Bay and Lavernock the range of the nominal species of the lowermost Jurassic *Ogmoconchella aspinata* Zone of Michelson (1975) *sensu* Boomer (1991*a*) extended down into Penarth Group beds. Other microfossil groups used for zonal schemes for the Upper Triassic, and with applications for the Penarth Group, include calcareous nannofossils and foraminifera. Hamilton (1982) discovered two nannofossil taxa in the Penarth Group, one of which, *Annulithus arkelli*, was proposed as a zonal indicator. Unfortunately, the range of this species spans the lengthy Rhaetian to middle Hettangian interval, and it is thus of limited value for fine scale biostratigraphical correlation. Foraminifera occur at several levels within the Penarth Group, and three diagnostic assemblages were delineated by Copestake (1989): a lower *Glomospira/Glomospirella* Assemblage extends from the upper part of the Blue Anchor Formation to the Cotham Member; an *Eoguttulina liassica* Assemblage occurs in the Langport Member, excluding the 'Watchet Beds'; and an assemblage termed the JF1 Zone commences in the highest parts of the Langport Member and continues into beds of the Blue Lias Formation of the Lias Group.

Finally, Swift (1989, 1995*a*, 1995*b*) documented and described conodonts recovered from beds of Rhaetian age in Britain, mainly from the Langport Member, and considered their biostratigraphical potential. The occurrence of the latest Triassic indicator *Misikella posthernsteini* in the Pre-planorbis Beds at Barnstone, Nottinghamshire allowed correlation to be made with a part of the latest Triassic *Rhabdoceras suessi* and *Choristoceras marshi* ammonite zones of the Alpine Tethyan region. This confirmed a latest Triassic age for the Preplanorbis Beds, and therefore demonstrated that the underlying Penarth Group must be older than latest Rhaetian.

3. BRIEF HISTORY OF PALAEONTOLOGICAL STUDIES

by ANDREW SWIFT

Establishing the earliest reference to fossils in the Penarth Group is difficult, but certainly one of the oldest records is that of E. Owen (1754), who gave an extensive description, including an illustration (see Frontispiece), of 'Cotham Stone' from Bristol (better known now as Cotham or Landscape Marble, see Chapter 17). Subsequent research on the palaeontology of the Penarth Group tends to have been split between palaeontologists studying the vertebrate fossils and others specializing in the invertebrates, with the flora forming the basis of a third, very much smaller, group of studies.

MACROFOSSILS

It was the vertebrate fossils of the bone beds of the Penarth Group and the partly contemporary fissure fillings in the palaeokarst of Carboniferous Limestone that first attracted the attention of British geologists (e.g. Buckland and Conybeare 1824; Riley and Stutchbury 1836; Owen 1840, 1842). Many of the fish taxa found in these deposits had been first described from European material by the famous Swiss palaeontologist Louis Agassiz (1833–44, 1844). The insects were also described at an early date in a large work by Brodie (1845), who also gave the first details of the liverwort *Naiadita* in the same publication.

The first serious attempt to document the invertebrate macrofauna was by Wright (1860). In this important paper, 13 species of bivalve were recognized and a single species of crustacean. In addition, Wright noted that reptilian remains were present in the bone beds of the Westbury Formation. Wright did not describe this fauna, but recognized that the bivalve *Avicula* (*=Rhaetavicula*) *contorta* Portlock was a characteristic and distinctive species which he nominated to define a biozone (the 'Zone of *Avicula contorta*') encompassing the lower part of the beds now known as the Penarth Group. His stratigraphical scheme excluded the higher beds i.e. the Langport Member of current usage, which he allocated to the Lias. Wright suggested that the fossils of the 'Zone of *Avicula contorta*' indicated correlation with the Koessen Beds of the eastern Alps of Austria in the type area of the Rhaetian.

Shortly after Wright's work, a paper of seminal importance by Moore (1861*b*) expanded considerably on the number of invertebrate species recognized from the 'Zone of *Avicula contorta*'. More than 50 species were listed and described, including the first records of a coral, a cirripede (barnacle), a number of echinoderm remains, worm traces and the first indisputable identification of a brachiopod. Notably, the descriptions included the teeth of the earliest mammal to be recovered from the Mesozoic of Britain ('*Microlestes*'). Many of Moore's original specimens have been re-figured for this guide. Moore (1861*b*) included the 'White Lias' (nearly everywhere the equivalent of the modern Langport Member) in the 'Zone of *Avicula contorta*', thereby bringing together for the first time the beds which now constitute the Penarth Group. Moore (1867) followed his earlier work with a long exposition on the Mesozoic geology of Somerset and South Wales, but no new taxa were described.

Another important paper of the 1860s by Dawkins (1864) described in detail the 'Rhaetic Beds and White Lias' of Somerset. The study consisted mainly of descriptions of exposures and lists of their fossils, but a new mammal taxon, the oldest recorded from Britain, was erected, based on teeth recovered from the 'Grey Marls' from the shore 'west of Watchet'. One of Dawkins' (1864) conclusions was that the fauna of the 'White Lias' was so distinctive as to exclude its allocation to either the 'Rhaetic' or the 'Lower Lias'. Pioneering studies of the vertebrate fossils of the fissure fillings and bone beds, the latter at Aust Cliff in particular, were running alongside the researches of Wright and co-workers and there was a number of short taxonomic papers published in the period 1860–90 (e.g. von Meyer 1867; Brodie 1872; Stoddart 1875; Sanders 1876; Davis 1881; A. S. Woodward 1889*a*, 1889*b*).

In the wake of the pioneering work of Agassiz, Owen, Wright, Moore, Dawkins and others, British geologists applied the new knowledge of the macrofauna of the Penarth Group to interpret the chief areas of outcrop of the group in Britain. Most workers included fossil lists or stratigraphical sections with fossil content appended (e.g. Wilson 1882; Brodie 1886). Usually, no new species were reported, but Harrison (1876) made the first report of an ophiuroid (brittle-star) and Reynolds and Vaughan (1904) erected two new species of bivalve. The latter authors also included detailed descriptions of both the vertebrate and invertebrate fossils from sections exposed by railway works near Chipping Sodbury. Tomes (1884, 1903) discussed and described the coral taxa of the Penarth Group first recognized by Duncan (1867–68), and Short (1904) attempted one of the first biozonations of Penarth Group with characteristic fossils nominated for each unit in sections in the Bristol district. Linsdall Richardson, arguably the greatest and certainly the most influential of Penarth Group stratigraphers, published a very large number of papers in the first third of this century. Amongst these were several seminal works (e.g. Richardson 1905, 1911) and fossil records were invariably included. Very occasionally Richardson described taxa (e.g. 1905), but he was not primarily a taxonomist.

The Geological Survey meanwhile was systematically mapping the country, and although taxonomic details were not included in their memoirs and other publications, their records of Penarth Group fossils were comprehensive and form invaluable databases (e.g. Jukes-Browne 1885; Woodward and Ussher 1906; Lamplugh *et al.* 1909). The major British museums, particularly The Natural History Museum, London, were also compiling catalogues which included Penarth Group biota, many of fundamental taxonomic importance (e.g. Lydekker 1888, 1889, 1890; A. S. Woodward 1889–1901; Allen 1899, 1904). Horwood (1913/16) gave a comprehensive fossil list for the Upper Triassic of Leicestershire, with most species recorded from the Penarth Group. Some taxonomic notes were included.

During the initial impetus of 'Rhaetic' research by geologists in the period 1860–1915, much of the basic stratigraphical and palaeontological work was completed and no major Penarth Group palaeontological research was instigated for the next 40 years, with the notable exception of Harris's meticulous description of the macroflora (1931, 1938, 1939). Also during this period Arkell (1933) completed his classic Jurassic memoir (which included the 'Rhaetic') and gave characteristic fossils for each unit in a tentative step towards a zonal scheme. After 1955, research into Penarth Group palaeontology increased. Harris (1957) and Harris and Jackson (1961) returned to the theme of the Penarth Group macroflora, while Robinson (1957) published the most comprehensive report to date on the vertebrates of the fissure fills of the Bristol Channel area. The presence of invertebrate palaeocommunities in the Langport Member of the south Devon coast was recognized by Hallam (1960), who described two distinct assemblages. The structures of Cotham or Landscape Marble, long mysterious, were recognized as being of algal origin by Hamilton (1961). Gardiner (1961) described new beetles from the Penarth Group, although the insects have remained a neglected group with potentially many taxa awaiting description (see Chapter 10). Cox (1961, 1963, 1964) published a series of papers which included revisions of some of the bivalves. The bone beds of the Westbury Formation and their distinctive vertebrate fauna again became the focus of study (e.g. Sykes 1971; Antia 1979), as they had been in the early years. Sykes *et al.* (1970) presented a comprehensive investigation of the vertebrate fauna of the main Westbury Formation bone bed at Barnstone, Nottinghamshire, together with an appraisal and description of the general macrofauna of the Penarth Group of that locality. Kent (1970) noted that elements of the fauna of the Preplanorbis Beds of the Lias Group in the East Midlands were present in the Langport Member of southern Britain, which suggested to him that locally the former may be a facies of the latter. Ivimey-Cook (1974), as part of a study of the Triassic of Wales, considered fossils and their distribution in the Penarth Group. Arthropod macrofossils, not reported since the last century, were described by Duffin (1978a) and Förster and Crane (1984). The Triassic-Jurassic boundary was the subject of a study by Poole (1979), who proposed that the boundary be taken at the top of the Cotham Member, since he believed that succeeding beds were characterized by a distinctive marine fauna with a marked Jurassic component.

The 1980s opened with a major review of the correlation of British Triassic deposits, wherein Warrington et al. (1980) considered the potential of the fossil data and concluded that the Penarth Group macrofauna, although indicative of a Late Triassic age, offered little scope for direct correlation between the British sequence and the 'Standard' type area in the eastern Alps. A new series of British Geological Survey memoirs covering classic areas of Penarth Group outcrop was published during this period and provided much of the basis of modern palaeontological and stratigraphical practice with regard to Penarth Group (e.g. Whittaker and Green 1983; Donovan and Kellaway 1984; Edmunds and Williams 1985; Waters and Lawrence 1987). The taxonomic lists in these publications record approximately 30 invertebrate genera in the British Rhaetian (which presently includes basal Lias Group beds that lack ammonites).

The momentum of palaeontological research begun in the 1980s continued through that decade and into the 1990s. The vertebrates in particular attracted attention and a new generation of palaeontologists addressed outstanding problems of their taxonomy and distribution (e.g. Fraser and Walkden 1983; Duffin 1985, 1994; Fraser 1985, 1986; Storrs 1993, 1994; Benton and Spencer 1995). In 1989 came the first record of an ammonite from the Penarth Group, a single, very small (3·8 mm diameter) form with psiloceratid affinities from the top of the Westbury Member near Chipping Sodbury (Donovan et al. 1989). Trace fossils, impressive and varied but hitherto largely overlooked, received long overdue attention by Wright and Benton (1987) and Wang (1993). Insects, also a long neglected group, were re-examined in part by Whalley (1982, 1983, 1985, 1988), but many still require major revision (see Chapter 10). The Triassic-Jurassic boundary debate was reopened (Hallam 1990, 1991; Cope 1991a, 1991b; Hodges 1994; Warrington et al. 1994) and fossil evidence was presented and refuted, with bivalves and the first appearance of ammonites prominent in the argument.

Warrington and Ivimey-Cook (1990) contributed a useful paper on the invertebrate fossils of the Upper Triassic and Lower Jurassic in the type sections in southern Britain, which summarized much of the data presented by earlier workers. First appearances, distributions and typical associations through the Penarth Group/Lias Group interval are given. The authors described the invertebrate macrofauna (excluding the insects) of the Penarth Group as consisting of gastropods, bivalves, a brachiopod, anthozoans, a cirripede, echinoid test and spine fragments, an ophiuroid and the single ammonite. Bivalves are the most numerous and diverse group, with Warrington and Ivimey-Cook (1990, p. 209) reporting 18 genera, three of which are restricted to the Westbury Formation (*Eotrapezium*, *Lyriomyophoria* and *Rhaetavicula*) and three found only in the Lilstock and younger formations (*Meleagrinella*, *Plagiostoma* and *Pleuromya*) (see Chapter 8). The same authors (Warrington and Ivimey-Cook 1995) documented in detail Penarth Group-Lias Group sections in west Somerset and South Glamorgan and gave comprehensive notes on the fossils of each unit, as well as their distribution.

Despite the presence of species of many fossil groups in the Penarth Group, biostratigraphy and correlation using macrofossils have not advanced significantly since Wright (1860) first registered the importance of *Rhaetavicula contorta*. This is still the only macrofossil commonly used to correlate British and Alpine sequences directly. Even this species allows only the correlation of the lower part of the Penarth Group, since it does not range beyond the basal beds of the Cotham Member of the Lilstock Formation. Other species are facies controlled or long-ranging, and the potential for further progress using the macrofauna seems small, unless more ammonites are discovered.

MICROFOSSILS

Ostracods from the Penarth Group were first reported by Duff (1842), who mistakenly thought that his faunas from Morayshire originated from Wealden deposits. Shortly afterwards, Brodie (1845) identified ostracods from undoubted Penarth Group horizons at Garden Cliff, near Westbury-on-Severn. Unfortunately, from this time a certain amount of confusion was generated, since Brodie apparently applied Penarth Group locality details to Wealden specimens. This resulted in the first systematic study (Jones 1894), which included Brodie's specimens, perpetuating these errors and describing Purbeck ostracods as originating from the Penarth Group. Jones (1894) described nine species (discounting varieties) from the Penarth Group and supposedly Lias Group horizons, but uncertainty attends the horizons from which certain taxa were recovered (see Anderson 1964, p. 133).

Anderson's (1964) paper was a major, but flawed, work and consisted of a study of existing collections and new material from two main sample sites and two boreholes, with some comparison with German occurrences. Anderson proposed an ostracod-based zonal scheme for the Penarth Group. Few later studies considered ostracods in any detail, but Kent (1968) referred to, and reproduced, Anderson's scheme in a review of the Penarth Group in the Midlands. Bate's (1978*a*) summary paper brought together all known ostracod data, and clarified some taxonomic problems. A range chart was given which integrated Anderson's (1964) zones. Brief notes on ostracods recovered from the Penarth Group at Pinhay Bay, Devon were included in a field guide by Lord and Bown (1987). Boomer (1991*b*, 1991*c*) revised the taxonomy of two Penarth Group species, whilst Lord and Boomer (1991) briefly considered ostracods with reference to the Triassic-Jurassic boundary. As a result of exploration for hydrocarbons in the Irish Sea, Ainsworth and Horton (1987), Ainsworth *et al*. (1987), Ainsworth (1989a, 1990), Ainsworth *et al*. (1989) and Rutherford and Ainsworth (1989) touched on aspects of ostracods from deposits which may be equivalent to some part of the Penarth Group.

The papers detailed above contain all the taxonomic information available on Penarth Group ostracods, but many other authors have noted ostracods in their faunal lists and section logs: e.g. Reynolds and Vaughan (1904, from the Cotham Member around Chipping Sodbury), Richardson (1905, from the Langport Member of Glamorgan), Edmunds *et al*. (1965, from the Cotham Member and Langport Member in Warwickshire), Ivimey-Cook (1974, from the Penarth Group at Lavernock Cove), Duffin (1980, from the Westbury Formation at Chilcompton) and Waters and Lawrence (1987, from the Penarth Group around Cardiff).

A small number of papers has been published on Penarth Group foraminifera and most are limited in scope both geographically and taxonomically. Chapman (1894) recognized and described a Penarth Group assemblage consisting of no less than 26 species from a quarry locality near Wedmore in Somerset. Many of these are of doubtful provenance and the whereabouts of Chapman's collections are unknown. Agglutinating forms predominated. There followed a long hiatus during which no further information on Penarth Group foraminifera was published. Barnard (1950), in a study devoted to the foraminifera of the Lias Group of the Dorset coast, noted that poorly preserved representatives of three genera were present in the Rhaetian–Hettangian strata of that area. Banner *et al*. (1971) studied submarine cores from the approaches to Barry, South Wales and also sampled coastal exposures of the Penarth Group nearby. Foraminifera were extracted from both the cores and onshore exposures and their biostratigraphical potential considered. Although most information was gleaned about Jurassic levels, the first undoubtedly Penarth Group occurrences of species of the genus *Geinitzinita* were reported. These species were later assigned to the genus *Lingulina* (Copestake 1989). A comparison of bivalve and foraminifera faunas from the Penarth Group of southern England and Wales with collections from broadly equivalent deposits in the classic Alpine areas was published by Hallam and El Shaarawy (1982). A marked disparity in faunas was found, which was attributed partly to salinity differences. Copestake (1989) is the most comprehensive publication to date and includes a summary of past work with the addition of a considerable amount of new data. An integrated approach was adopted and for the first time diagnostic assemblages were recognized and delineated for the units of the Penarth Group. Few other references to foraminifera are available and these are limited to records of occurrences only (e.g. Edmunds *et al.* 1965; Ivimey-Cook 1974; Waters and Lawrence 1987).

Conodonts, in view of the recent discovery of whole body fossils (Briggs *et al.* 1983), are now better considered as macrofossils, but the traditional treatment of their phosphatic elements as microfossils is followed here. These important biostratigraphical indicators were discovered only recently in the British Rhaetian (Swift 1989), and proved to be extremely small and fragile. Among the collections was an element from the apparatus of *Misikella posthernsteini*, an indicator of the uppermost Triassic (Kozur 1989). This was recovered, not from the Penarth Group, but from a horizon in the basal Triassic part (Preplanorbis Beds) of the Lias Group at Barnstone, Nottinghamshire. This occurrence, in beds overlying the Penarth Group, demonstrates that the group is wholly Triassic and allows direct correlation with many Late Triassic sequences around the world, including the type area in the Austrian Alps. Swift (1995*b*) followed his earlier publication with a comprehensive systematic analysis of all known collections of British Late Triassic conodonts.

4. ASPECTS OF THE SEDIMENTOLOGY OF THE WESTBURY FORMATION

by JOE H. S. MACQUAKER

The striking visual stratigraphical setting of the Westbury Formation and the unusual occurrence within it of beds rich in vertebrate remains has drawn many geologists to investigate its origin (e.g. Richardson 1933; Warrington and Ivimey-Cook 1995; Martill this volume). Given the current economic and academic interest in 'black shales' (Wignall 1994), a consideration of aspects of the sedimentology of the Westbury Formation sheds revealing light on both the palaeontology and environmental setting of a distinctive and exceptional group of rocks.

The formation is composed predominantly of siliciclastic-rich mudstones with subordinate interbedded calcareous sandstones, bioclastic packstones ('shell-beds'), wackestones and intraformational conglomerates. It is commonly interpreted as comprising a succession of upward-fining units ('cycles', see Warrington and Ivimey Cook 1995), which are most commonly described as having either 'bone beds' sandstones or bioclastic packstones at their bases and mudstones towards their tops. The formation was deposited within a series of fault-bounded basins (e.g. the Worcester Graben, Bristol Channel Trough, Wessex Basin) on the north-west margin of Tethys (e.g. Ziegler 1982). In spite of the fact that many of its mudstones contain organic matter, the Westbury Formation is a poor source rock at outcrop, having undergone insufficient burial to generate oil (Macquaker *et al.* 1986; Tuweni and Tyson 1994).

The sediments of the Westbury Formation are most commonly interpreted as having been deposited in a shallow marine setting in a storm-dominated epeiric sea (e.g. Macquaker 1994; Warrington and Ivimey-Cook 1995), a model supported by the faunal and biomarker contents. The 'bone beds' and intraformational conglomerates have been variously interpreted as transgressive lag deposits, condensed horizons or storm deposits. The calcareous sandstones have been considered to represent barrier bar deposits, whilst the bioclastic packstones represent winnowed shallow marine concentrations. Intervening mudstones are either offshore or lagoonal deposits (e.g. Mayall 1981; Macquaker 1994; Warrington and Ivimey-Cook 1995).

LITHOFACIES TYPES

Clay- and silt-rich mudstones (Pl. 7, figs 1–4)
Most of the Westbury Formation comprises clay- and silt-rich mudstones,

PLATE 7

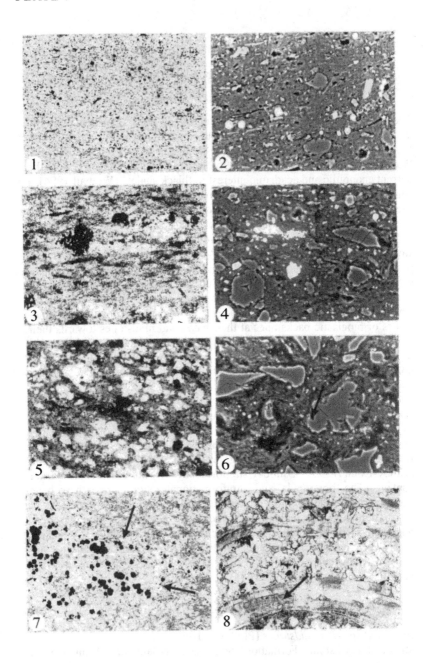

which have been described variously as 'black shales', 'fossiliferous shales' or 'paper shales'. Detailed petrological examination reveals that they are unlaminated and predominantly composed of clay with minor silt, organic matter and pyrite. The clay in these units is composed of mixed layer illite/smectite, with minor kaolinite, chlorite and glauconite. The silt component, which is variable (>5 per cent. in the silt-rich mudstones and <5 per cent. in the clay-rich mudstones), is predominantly composed of quartz with minor K-feldspar. The visible organic matter comprises mainly amorphous organic matter with minor algal and higher terrestrial plant material, with the bitumen fraction derived mainly from marine organisms (Macquaker *et al.* 1986; Thomas *et al.* 1993). Thin shell pavements consisting of disarticulated valves of *Rhaetavicula* sp., *Isocyprina* (*Eotrapezium*) sp. and *Chlamys* sp. are present at some levels.

Sand-rich mudstones and muddy sandstones (Pl. 7, figs 5–6)
A few sand-rich mudstones and muddy sandstones are present. These are not laminated and are composed predominantly of clay (mixtures of illite and smectite) and pyrite with some silt and fine sand (predominantly quartz with minor K-feldspar). In addition, they contain glauconite and

---

EXPLANATION OF PLATE 7

Optical and backscattered electron micrograph pairs illustrating mudstone lithofacies present in the Westbury Formation.

Figs 1–2. Clay-rich mudstone from Lavernock. The sample is predominantly composed of clay with minor silt (low η), pyrite (high η) and organic matter (very low η). Optical micrograph field of view 700 μm; BSE micrograph field of view 150 μm.

Figs 3–4. Silt-rich mudstone from Watchet Harbour. The sample is predominantly composed of clay with some silt (comprising quartz and K-feldspar) with minor pyrite. Note that the pyrite has a framboidal texture. Optical micrograph field of view 700 μm; BSE micrograph field of view 150 μm.

Figs 5–6. Sand-rich mudstone from Watchet Harbour. This sample is predominantly composed of clay and amorphous organic matter with some silt (comprising etched K-feldspar see arrow on 6) and minor pyrite. Optical micrograph field of view 700 μm; BSE micrograph field of view 150 μm.

Fig. 7. Muddy sandstone from Watchet. This sample is composed of fine sand-sized quartz and K-feldspar in a matrix of clay and amorphous organic matter. In this sample the sand infills small burrows (margin arrowed) and forms starved ripple laminae. In the sand-dominated regions the intergranular porosity is infilled by calcite and pyrite. Field of view 1·7 mm.

Fig. 8. Calcareous sandstone from Watchet Harbour. This sample is predominantly composed of medium sand-sized quartz and K-feldspar. In addition it contains some disarticulated bivalve fragments (which exhibit pseudopleochroic fabrics, arrowed) and is cemented by non-ferroan calcite and pyrite. The sample has a high minus cement porosity. Field of view 1·7 mm.

organic matter. At some locations they are burrow mottled (*Chondrites* isp.), contain starved sand ripples and may contain broken shell debris. Where the sand is abundant it is commonly cemented by non-ferroan calcite (see below).

Calcareous sandstones (Pl. 7, figs 7–8)
Minor thin calcareous sandstones are also present in the Westbury Formation. These are commonly either ripple laminated or intensely bio-turbated, usually exhibiting exquisite trace fossils (e.g. ichnospecies of *Arenicolites*, *Planolites*, *Cruziana*, *Rusophycus*, *Diplocraterion*, etc.; see Wright and Benton 1987; Wang 1993; Chapter 16) and form both continuous horizons and discontinuous, starved-ripple-laminated units. Internally, some ripples exhibit upward bundle textures (cf. Raaf *et al.* 1977), some have collapsed ripple crests with trapped mud drapes (cf. Aigner 1982), whilst others may form gutter casts (cf. Rawson *et al.* 1982). Disarticulated shells (e.g. *Modiolus* sp.) and vertebrate-derived phosphatic debris commonly occurs within the ripple troughs. The cal-careous sandstones are composed predominantly of fine sand-sized quartz and K-feldspar, have a variable reworked vertebrate derived component and were cemented early by either non-ferroan calcite and/or pyrite.

Wackestones (Text-fig. 4A–B)
Within certain horizons thin nodular wackestones are developed. These either form continuous layers or discontinuous concretions. They are composed mainly of microcrystalline calcite cements and may contain uncompacted shell debris (e.g. *Chlamys* sp.) and pyrite. These wacke-stones exhibit high minus cement porosities and the enclosing mudstones are commonly compacted around them. Some of the bioclastic mudstones exhibit septarian cracks.

Bioclastic packstones (Text-fig. 4C–D)
Thin bioclastic packstones ('shell-beds' *sensu* Kidwell *et al.* 1986) are a minor component in the Westbury Formation. They are predominantly composed of disarticulated, broken shell debris (e.g. *Modiolus* sp., *Chlamys* sp.), have a minor sand component, are cemented by carbonate cements (e.g. non-ferroan and ferroan calcite) and have high minus cement porosities. The shell debris, which is commonly micritized, is preserved mainly convex-up although some is preserved 'end-on' (*sensu* Kresia and Bambach 1982) and may be current-oriented (e.g. Kelling and Moshriff 1977). The shelter porosity, which has been commonly infilled by a combination of microsparry calcite and by a number of zoned calcite cements (non-ferroan and ferroan calcite), is preserved beneath the convex-up shell fragments. In some units superficial ooids are present. In these the grain coats either enclose shell debris or detrital quartz grains. At

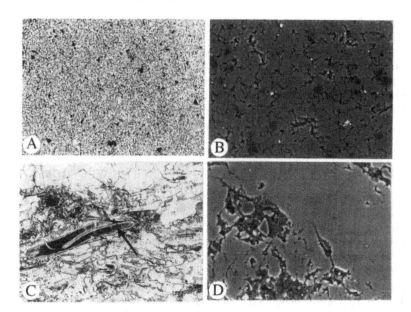

TEXT-FIG. 4. Optical and backscattered electron micrograph pairs which illustrate a typical wackestone and bioclastic packstone from the Westbury Formation. A–B, wackestone from Watchet Harbour. The sample is predominantly composed of microcrystalline calcite (intermediate η) with minor dolomite (low η) and trace pyrite. Note the high minus cement porosity. Optical micrograph field of view 700 μm; BSE micrograph field of view 150 μm. C–D, bioclastic packstone from Watchet Harbour. The sample is predominantly composed of recrystallized, calcarous bivalve debris with minor, phosphatic, vertebrate-derived material (arrowed). The intergranular porosity is infilled by a fabric-destructive fibrous calcite cement. Optical micrograph field of view 700 μm; BSE micrograph field of view 150 μm.

some localities (e.g. St Mary's Well Bay) the bioclastic packstones have been pervasively recrystallized. In these units the shell fragments either preserve pseudopleochroic textures or are only recognizable where they are outlined by pyritized micrite envelopes (e.g. at Aust and Wainlode cliffs). In addition to micritization, at some localities the shell debris has been extensively bored, as at Watchet Harbour (Macquaker 1994).

Intraformational conglomerates
In addition to the mudstones, sandstones and limestones, the Westbury Formation also contains some intraformational conglomerates (e.g. the 'bone beds' at Aust and Garden cliffs, Watchet and Lavernock Point; see

Chapter 5). These commonly contain intraclasts up to 300 mm in diameter derived either from the underlying Blue Anchor Formation and/or from within the Westbury Formation. At some localities the clasts are derived from broken pieces of desiccation-cracked mudstone (Warrington and Ivimey-Cook 1995) and these units may be classed as flat pebble conglomerates. At some localities (e.g. Aust Cliff, Garden Cliff and Lilstock) the intraformational conglomerates occur at the base of the formation whereas at others (e.g. Lavernock and Watchet) they occur within the succession.

LITHOFACIES STACKING PATTERNS IN THE WESTBURY FORMATION

Successive beds in the Westbury Formation either coarsen or fine upwards. The details of this lithofacies stacking are complex and are developed on at least two scales. For instance, within an individual cemented unit (i.e. on an individual bed scale) a number of facies may be present. Here bioclastic packstones may alternate with calcareous sandstones, wackestones and intraformational conglomerates to form composite beds comprising alternating upward-fining facies couplets separated by erosive events. On a larger scale (i.e. on a stacked bed scale), both upward-fining and upward-coarsening can be observed. At this scale upward-coarsening manifests itself in terms of successive units having increasing proportions of silt and sand with clay-, silt- and sand-rich mudstones progressively giving way upwards to calcareous sandstones and bioclastic packstones. Conversely, upward-fining manifests itself in terms of successive units having decreasing proportions of silt and sand with sand- and silt-rich mudstones gradually giving way upwards to bioclastic wackestones. At this larger scale the cemented units commonly occur at the levels where the stacking patterns change. Moreover, relative to the rest of the Westbury Formation sedimentary rocks, the coarser units at the top of the larger scale upward-coarsening successions show most evidence of current activity (ripple lamination, reworking) and prolonged macrofaunal colonization (as indicated by the presence of trace fossil tiering – N.B. here both *Thalassanoides* isp. and *Chondrites* isp. can be observed within individual units).

The number of upward-coarsening and fining successions in the Westbury Formation varies from locality to locality. Those in the Worcester Graben commonly have from four to six units whereas successions in the Bristol Channel Trough usually have more than six.

DEPOSITIONAL AND EARLY DIAGENETIC ENVIRONMENTS

The combination of a diverse lithofacies assemblage and a wide variety of sedimentary structures suggests that the depositional environment of the Westbury Formation was variable. Moreover, as this variability is present both on the scale of individual beds and on a stacked-bed scale it is likely that the processes which caused these differences occurred on a variety of scales ranging from individual events (e.g. storms) occurring on a human time scale to progressive climate/relative sea-level changes on a much longer time scale (10×10^3 to 10×10^5 years).

On a short time scale the presence of lithofacies couplets, symmetrical ripple laminae, spill over ripples, gutter casts, intraformational clasts and shell debris preserved in 'end-on' orientations within the calcareous sandstones and bioclastic packstones suggest that storms reworked the sediment (see Kresia 1981; Aigner 1982; Rawson *et al.* 1982; Seilacher 1982; Kidwell 1989; Lesuer and Tastet 1994; Macquaker 1994).

On a longer time scale, the varying lithofacies suggest that the sediment supply was changing. In this context it is likely that the upward-coarsening successions were deposited in environments that were gradually shoaling upward and the upward-fining successions were deposited in settings where they were gradually foundering. These long term differences suggest that during deposition of the Westbury Formation accommodation space, rather than being fixed, was variable. In the absence of any obvious local tectonic cause for these patterns it is likely that there was either relative sea-level change and/or the sediment supply into the basins was progressively changing. Given that similar cyclicity can be observed in all the depositional basins, and that there is evidence of limited sub-aerial exposure at the top of at least one of the coarsening upward successions in the Bristol Channel Trough (mudcracked horizon at Watchet, Blue Anchor Bay and Lavernock) the possibility of relative sea-level changes as the controlling mechanism is most likely.

'Shell beds' are commonly observed at the tops of upward shoaling successions in stratigraphical settings where there is very little accommodation space (*sensu* Jervey 1988), high energy (e.g. Kidwell 1989, 1991) and sediment bypass (Macquaker and Taylor 1996). Given that the shell beds in the Westbury Formation commonly occur in a similar stratigraphical setting, contain superficial ooids and a diverse, tiered community of trace fossils, it is likely that they too formed under similar conditions. Moreover, as the shell beds and calcareous sandstones are preferentially cemented and exhibit high minus cement porosities, it is likely that they precipitated prior to compaction close to the sediment-water interface during periods when there were significant breaks in sedimentation. The presence of zoned non-ferroan calcite, pyrite and ferroan calcite cements in these units suggests that cementation was caused by anaerobic bacterial degradation of organic matter in the sulphate reduction and methanogenic zones (e.g. Claypool and Kaplan 1974; Irwin *et al.* 1977; Curtis 1980, 1987).

The units which contain abundant vertebrate derived debris ('bone beds') are best developed in stratigraphical settings over which there is rapid deepening as indicated by the presence of cemented transgressive lag deposits overlain by mudstones (Macquaker 1994). These occur either on the top of shell-beds/calcareous sandstones or at the major flooding event at the base of the Westbury Formation above the Blue Anchor Formation, as at Aust and Lavernock. The intervening mudstones may contain up to 5 per cent. organic matter, but are not laminated. This suggests that they were deposited in regions where the bottom waters were oxic. This interpretation is reinforced by the presence of occasional shell pavements and burrow mottling. Both of these attributes suggest that there was sufficient oxygen in the bottom waters to support an aerobic community during deposition of the mudstones. On the other hand, the presence of pyrite, organic matter and characteristic cements associated with sulphate reducing bacteria (i.e. pyrite and non-ferroan calcite), suggests that once buried the pore waters in these units became anoxic and reducing (e.g. Berner 1984).

Significantly, bioclastic wackestones are present at the levels where the stacking patterns change from overall upward-fining to upward-coarsening. Their composition (non-ferroan calcite), texture (nodular, microsparry cement which pre-dates compaction) and bedding parallel relationships suggest that they too precipitated close to (within a metre of) the sediment/water interface as a result of anaerobic bacterial decay processes in response to breaks in sediment accumulation (Curtis 1980; Macquaker 1994). Their stratigraphical setting, however, indicates that their precipitation was not linked to a break in sediment accumulation associated with a period of sediment bypass, as is the case for the cementation of the shell beds and calcareous sandstones (see above), but rather that it was related to periods of very slow sedimentation linked to reduced clastic input (i.e. at condensed sections *sensu* Loutit *et al.* 1988).

STRATIGRAPHICAL INTERPRETATION OF FACIES IN THE WESTBURY FORMATION

The temporal variations in sediment supply in the Westbury Formation suggest that these sediments can be interpreted in terms of sequence stratigraphical principles (*sensu* Van Wagoner *et al.* 1990). The presence of both upward-coarsening and upward-fining successions suggests that the large scale stacking patterns are developed on a stacked parasequence scale. In this context the upward-coarsening successions are interpreted to be retrogradational parasequence sets, the upward-fining successions are interpreted to be progradational parasequence sets and the cemented units, at the levels where the stacking patterns change, are interpreted to be equivalent to the main stratal surfaces. With reference to the key stratal surfaces it is reasonable to argue that the bioclastic wackestones, which occur at the level where the stacking patterns change from upward-fining to upward-coarsening, precipitated in response to a break in sediment accumulation at the maximum flooding surfaces (MFS). Unfortunately, it is much more difficult to interpret to which stratal surface the calcareous sandstones and bioclastic packstones are related. It is tempting to assume that, as they are best developed at the levels where the stacking patterns change from overall upward-coarsening to upward-fining, their formation was linked to sequence boundary (SB) development. In the Westbury Formation, however, it is not always possible to identify unequivocally the sequence boundary as commonly there is very little evidence of subaerial exposure. Additionally, in stratigraphical settings such as these, in theory, the change in stacking patterns at this level could mark a stratal surface between a forced regressive system tract (FRST) (*sensu* Hunt and Tucker 1992) and an overlying transgressive system tract (TST) rather than between the high stand system tract (HST) and the TST. If true, then this surface would occur above the classical definition of a sequence boundary (*sensu* Van Wagoner *et al.* 1990). Thus, it seems prudent to interpret the progradational parasequence sets in the Westbury Formation to be combined HST/FRSTs where evidence of subaerial exposure is lacking, and as HSTs where flat pebble conglomerates mark the SB (e.g. Watchet and Lavernock). Logically, it is reasonable to interpret the retrogradational parasequence sets to be TSTs, and the calcareous sandstones and bioclastic packstones to have formed within sequence boundary zones. The nodular wackestones are interpreted to have precipitated in the sediment below condensed sections (see also Macquaker and Taylor 1996) at MFSs.

Finally, given the historical interest in 'bone bed' formation, it is worth considering their stratigraphical significance in the Westbury Formation. Using their stratigraphical setting as a guide, i.e. their association with transgressive lag deposits, their presence immediately underlying major flooding surfaces and either capping upward-coarsening successions or unconformably overlying the Blue Anchor Formation, it is most probable that beds rich in vertebrate debris mark coincident sequence boundaries/flooding surfaces (e.g. Macquaker 1994; Macquaker *et al.* 1996).

CONCLUSIONS

Sedimentological analyses using combined optical, electron optical and field descriptions indicate that the Westbury Formation comprises: clay-rich mudstones, silt-rich mudstones, sand-rich mudstones, muddy sandstones, calcareous sandstones, bioclastic packstones and wackestones. These units contain a marine fauna and flora and were predominantly deposited subaqueously on a shallow, storm-dominated marine shelf.

Upward-coarsening and upward-fining successions on a stacked parasequence scale within the Westbury Formation suggest that sediment distributions were controlled by relative sea-level changes. Upward-coarsening successions are interpreted as HSTs where the sequence boundaries are unequivocally identifiable, and as combined HSTs/FRSTs where they are not. In this succession the upward-fining successions are interpreted to be TSTs, the levels where the stacking patterns change from overall upward-fining to upward-coarsening (marked by wackestones) are interpreted to be MFSs, the levels where the stacking patterns change from overall upward-coarsening to upward-fining (marked by calcareous sandstones and bioclastic packstones) are interpreted to be sequence boundary zones and the levels at locations where intraformational conglomerate horizons are present are interpreted to be the sequence boundaries. Finally, the 'bone beds' in the Westbury Formation are interpreted to have formed on surfaces where sequence boundaries and flooding surfaces are coincident.

5. BONE BEDS OF THE WESTBURY FORMATION

by DAVID M. MARTILL

A characteristic feature of the lower part of the Westbury Formation of the Penarth Group is the presence over most of the outcrop of thin accumulations of coarse clastic material, often associated with abundant coprolites, fish teeth and scales, and occasional bones of larger marine and, more rarely, terrestrial reptiles (Text-figs 5–7; Pl. 8). These deposits are known widely as the 'Rhaetic bone beds', despite bone often being only a minor constituent. Some horizons have received local names, such as the 'Basal Bone Bed' for bone beds in the Severn Valley region which overlie the Blue Anchor Formation and 'The Clough' in South Wales (Richardson 1911).

There have been numerous studies of the bone beds, partly because of the diversity of the contained vertebrate assemblage but also because they are clearly unusual deposits recording conditions of sedimentation contrasting strongly with those of associated strata above and below. Perhaps

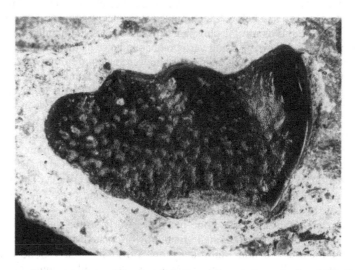

TEXT-FIG. 5. Dental plate of the lungfish *Ceratodus* sp. from the basal Westbury Formation bone bed at Aust Cliff, Gloucestershire. An example of a large skeletal element of a freshwater species found in association with skeletal elements of marine reptiles; University of Leicester, Department of Geology specimen LEIUG 49469; *c.* ×1·5.

the most remarkable feature of the bone beds is that despite variability both in clast size, faunal content and overall character, they can be observed at approximately similar horizons at points along an outcrop more than 400 km long. This strongly suggests that the genesis of bone beds in the Westbury Formation is partly time dependant and is most likely linked to the first major phase of the Rhaetian marine transgression, rather than to 'local' conditions.

The origin of the bone beds has been discussed by many workers (e.g. Short 1904; Wickes 1904; Reynolds 1946; Macquaker 1994) and some of the models proposed to explain the genesis of bone beds as well as aspects of bone bed distribution, sedimentology and taphonomic palaeontology are considered below. Discussions of the individual taxa occurring within the bone beds are to be found under the relevant systematic chapters herein, chiefly chapters 14 and 15.

THE TERM BONE BED

A number of workers have attempted to define bone beds using criteria such as amount of vertebrate skeletal element content and phosphate concentration (e.g. Reif 1976, Antia 1979), but no single satisfactory scheme has emerged and the term bone bed is generally used to refer to those deposits which contain more abundant vertebrate elements than related beds above and below. Using phosphate concentration as a criterion can be misleading, as not all phosphate in sediments is contained within bones and teeth. Likewise, using vertebrate abundance as the sole criterion can also be less than useful. Some vertebrate remains are often rare, but over time, collecting can produce a diverse vertebrate fauna from a particular horizon. The term is sometimes used by palaeontologists to describe informally a horizon which has yielded a single skeleton, even when there are no other associated vertebrates. Here, it is suggested that any deposit in which multiple vertebrate skeletal elements (e.g. teeth, scales, bones) can be seen in hand specimen, can appropriately be termed a 'bone bed'. Sykes (1977) defined four types of bone bed in the Penarth Group using a combination of sedimentological and palaeontological criteria. The four types are: (1) primary bone beds; (2) secondary bone beds; (3) scatter bone beds; and (4) trace bone beds.

Characteristically, primary bone beds are poorly sorted, lack bedding (or are only poorly bedded), contain randomly orientated, articulated shells and contain long, slender unbroken coprolites. Secondary bone beds typically are well-sorted, show normally graded bedding and contain fossils that show evidence of considerable transport and abrasion. Scatter bone beds are usually sandy or silty mudstones with sparse vertebrate remains disseminated throughout the bed. Trace bone beds are very thin layers with patches of accumulated bones, usually only a single layer thick.

According to Sykes (1977), all other bone beds can pass laterally into this latter type of bone bed.

STRATIGRAPHICAL DISTRIBUTION

Bone beds in the Penarth Group occur most commonly towards or at the base of the Westbury Formation and take the form of conglomerates or sandstones containing numerous, often abraded, small vertebrate remains. Where pyrite or calcite is present as cements the bone beds can be exceedingly hard and produce prominent ledges on cliff faces. The coarser (major) bone beds usually rest directly and disconformably on the eroded top of the Blue Anchor Formation or in some places with marked unconformity on Carboniferous strata (Curtis 1981). At Blue Anchor Point, Somerset and at Culverhole, Devon, a gritty, basal bone bed lies on an eroded surface of the Blue Anchor Formation and fills burrows in its surface (C. Duffin, pers. comm.). Other, generally finer-grained bone beds may occur within the basal part of the Westbury Formation, and several bone beds may be present at a single locality (e.g. four bone beds have been documented from many localities in the south-west and, exception-ally, eight, at Barrow upon Soar, Leicestershire (Sykes 1977)). These usually occur within the lowest 2–3 m of the Westbury Formation, although thin accumulations of sand or silt with bones have been found higher in the sequence (Sykes 1977). At Penarth, *Thalassinoides* burrows within the Westbury Formation are filled with arenaceous bone-bearing sediment (C. Duffin, pers. comm.). Macquaker (1994) noted that the bone beds often occur at the top of coarsening upward cycles but many of the individual bone beds have very sharp, erosive bases, and are thus part of new sedimentation events (Hamilton 1962; Duffin 1980). Many of the conglomeratic bone beds grade upwards into finer-grained bone-bearing sands and silts, and as such, are parts of fining-upward sequences. This can be explained most easily if their genesis is considered in terms of the progressive shoreward reworking of pre-existing bone-bearing sediments that are buried offshore.

GEOGRAPHICAL DISTRIBUTION

The best documented and most readily accessible of the Westbury Formation bone beds occur in the north Somerset coastal outcrops, South Wales and the shores of the River Severn at Aust, Westbury, Garden and Wainlode cliffs. Inland, bone beds have been documented from the Mendips (Duffin 1980), from long-lost sites in Leicestershire at Glen Parva and Spinney Hills (Harrison 1876; Wignall *et al.* 1989) and in active gypsum mines in Nottinghamshire at Newark (Martill and Dawn 1986), and Stanton on the Wolds (Wilson 1882). Perhaps the best documented of the inland sections was previously exposed in the disused Barnstone railway cutting, Nottinghamshire (Sykes 1971, 1974a, 1974b, 1977, 1979). Sykes (1977) also gave stratigraphical sections for many other localities where bone beds can still be examined, as well as reviewing their stratigraphical distribution in sites no longer accessible. In boreholes, pyritic silty shales yielding fish fragments and coprolites have been recorded at Felixkirk in North Yorkshire (Ivimey-Cook and Powell 1991). Vertebrate-bearing horizons have also been noted in outcrops of the Westbury Formation in Northern Ireland (Wilson 1981) and on the island of Mull (Lee and Bailey 1925). On the east coast of Scotland vertebrate-bearing sediments occur within possible Rhaetian deposits in the Linksfield erratic, near Elgin, Morayshire (Taylor and Cruickshank 1993) but these are not a bone bed facies.

At any one outcrop the bone beds are usually laterally discontinuous and in the case of basal bone beds, appear to be filling very shallow and almost imperceptible hollows in the top of the Blue Anchor Formation. At Bantycock Quarry, Newark, Nottinghamshire, the top of the Blue Anchor Formation is fissured, channelled and extremely uneven. Here a basal vertebrate-rich sandstone of the Westbury Formation penetrates these fissures to a depth of up to 2 m (Martill and Dawn 1986; A. Swift, pers. comm.).

SEDIMENTOLOGY

The sedimentology of the Westbury Formation bone beds is extremely variable considering their relatively confined temporal occurrences. This is largely a result of variability in the local availability and composition of input material (bioclasts and detritals), rather than major differences in the mode of bone bed genesis (Pl. 8, fig. 4). Bone beds lying at the contact with the underlying Blue Anchor Formation have sharp, erosive contacts and are coarsely conglomeratic in many places, containing clasts of up to 300 mm diameter derived from the underlying Blue Anchor Formation (Text-fig. 6). At Aust Cliff, Gloucestershire, a conglomeratic bone bed is particularly well developed, but lies a few tens of millimetres above the base of the Westbury Formation. Here, clasts of silty mudstones from the Blue Anchor Formation are variably rounded from subangular to subrounded. Hamilton (1977) noted that some of the Blue Anchor Formation mudstone clasts are deformed and suggests that they must have been partially lithified/plastic when incorporated into the bone bed. The larger clasts rest directly on the surface of the underlying Blue Anchor Formation but are often separate from each other. In section, this gives the appearance of a matrix-supported conglomerate, but this is clearly not the case. Sediment between the larger clasts is fine, angular quartz sand with common larger clasts. In the basal bone bed at Aust Cliff, quartz grains are

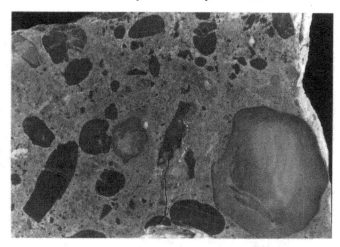

TEXT-FIG. 6. Cut surface of conglomeratic bone bed from Aust Cliff, Gloucestershire. Large subrounded clasts of mudstone occur with coprolites and bone debris in a groundmass of grit and sand cemented by calcite. Note that the largest coprolite (bottom left) has a broken end indicating prefossilization and its derived nature. University of Portsmouth collection; *c.* ×0·75.

angular and irregular, but a few euhedral crystals occur also (Antia 1979). Elsewhere, however, Antia and Sykes (1979) reported that quartz grains are often well rounded. Rarer sand size detrital particles include epidote, muscovite mica and lithic fragments. Hamilton (1977) has suggested that some of the larger, well-rounded quartz pebbles that occur in the bone bed at Aust Cliff might have been introduced into the basin as stomach stones of plesiosaurs. At Spinney Hills, Leicestershire, occasional pebbles in the thin (100 mm) basal bone bed are thought to have been derived from the Precambrian rocks of Charnwood Forest (Wignall *et al.* 1989).

Bioclasts occur within the interstitial sand /silt and comprise scales and teeth of fish. These clasts are of widely varying sizes, from less than 1 mm to several tens of mm in diameter (Pl. 8, figs 1, 4–5, 7). Broken bones and bone shards also occur and more rarely shelly fossils. Fining upwards can frequently be observed in the upper half of the bone bed with fine scale lamination visible in the highest parts of the bed, the top of which may also be ripple marked. The basal bone bed at Chilcompton is ripple laminated and has been described in detail by Duffin (1980).

Some examples of bone bed are lenticular and resemble narrow channel fills (flat topped lenses of between 100 to 1000 mm wide, with maximum thickness of between 100 and 200 mm). These show graded bedding, are very coarse at their bases with mudstone clasts and may show slump

PLATE 8

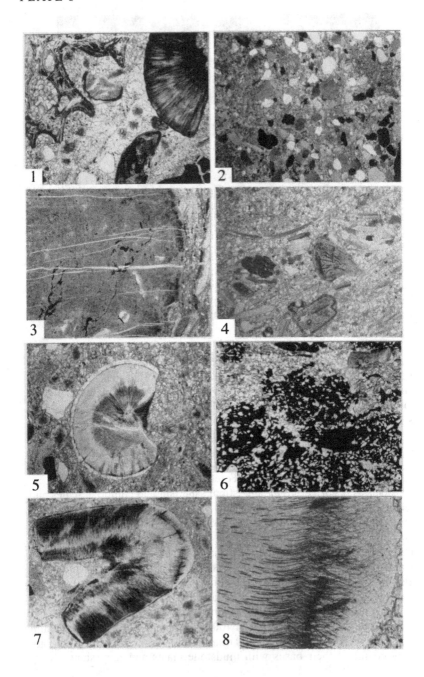

structures. Such bone beds possibly formed as sub-marine channel fills in front of river mouths. Small scale channel fill structures were described as early as the last century for Westbury Formation bone beds at Goldcliff, Monmouthshire (Lee 1872). Similar structures are found in bone beds from the Vectis Formation (Lower Cretaceous) of the Isle of Wight.

GEOCHEMISTRY

Few studies have been undertaken on the geochemistry of bone beds, and only the study of Trueman and Benton (1997) is of importance for the discussion on the origin of the Rhaetian bone beds of the UK. These workers analysed the trace element compositions of bones and matrix from bone beds at Aust Cliff near Bristol and Garden Cliff at Westbury-on-Severn. They were able to detect significant differences in the trace element composition between the bones and the matrix at Aust Cliff, but at Garden Cliff the matrix and bones had similar trace element compositions. Furthermore, the trace element composition of bones at Aust Cliff matched very closely that of the bones and matrix of Garden Cliff. Trueman and Benton (1997) interpreted this to mean that the source of the bones for both Aust Cliff and Garden Cliff was the same, or had very

---

EXPLANATION OF PLATE 8

Thin sections through examples of Westbury Formation bone beds highlighting sedimentological, taphonomic and diagenetic features.

Fig. 1. Conglomeratic bone bed with clasts of fragmented bone (upper left), broken teeth (upper right) in matrix of coarse angular quartz sand cemented by calcite; photographed under crossed polarized light; Aust Cliff, Gloucestershire.

Fig. 2. Bone sand, a matrix of angular to sub rounded medium quartz with bones and teeth cemented by calcite; Wainlode Cliff, Gloucestershire.

Fig. 3. Section through coprolite showing late cracking that persists into surrounding matrix; Aust Cliff, Gloucestershire.

Fig. 4. Shelly bone bed; disarticulated, thin valves of bivalves associated with broken shell debris with broken bones and teeth; Aust Cliff, Gloucestershire.

Fig. 5. Bone bed with large bioclasts in fine quartz sand cemented by calcite. The large circular object is a cross section through a fish tooth showing a bright outer enamel layer and a central zone of dentine; Aust Cliff, Gloucestershire.

Fig. 6. Fine bone sand cemented by calcite and pyrite (black); Wainlode Cliff, Gloucestershire.

Fig. 7. Tooth of the fish cf. *Sargodon* sp. with dentine tubules filled with pyrite. The pyrite is restricted to the tooth suggesting that it is reworked; Aust Cliff, Gloucestershire.

Fig. 8. High magnification photomicrograph of the tooth in fig. 7, at junction between dentine and enamel cap; dentine tubules clearly outlined by the pyrite infilling; *c.* ×70.

All *c.*×12, except where indicated.

similar early diagenetic environments, prior to being reworked and incorporated into the Westbury Formation bone beds. They further suggested that the Aust Cliff bone bed material might be derived from a bed equivalent to the Garden Cliff bone bed. Sedimentological evidence supports this, or a similar scenario, in which all of the vertebrate material in the Westbury Formation is derived. The study of Trueman and Benton (1997) was a preliminary analysis, and additional work on the trace element compositions between different taxa may also prove fruitful. For example, it would be interesting to make comparisons between freshwater fishes, marine fishes, marine reptiles and terrestrial reptiles.

A study of ^{87}Sr/^{86}Sr ratios by Schmitz *et al.* (1991) in biogenic phosphates of the fish *Saurichthys acuminatus* from the Westbury Formation bone beds showed a considerable discrepancy from those expected for Rhaetian sea water from other evidence. These authors concluded that the bone bed material may have been reworked or was from a freshwater deposit. These conclusions lend support to the findings of Hallam and El Shaarawy (1982) who suggested that a reduction in salinity in the Alpine region extended into north-western Europe in the latest Triassic. Clearly, integrated geochemical studies may help to elucidate the many interesting problems that remain.

FAUNA OF THE BONE BEDS

Both vertebrate and invertebrate macrofossils occur in the bone beds (Duffin 1978*b*; Storrs 1994). Microfossils recovered include both calcareous and phosphatic forms as well as palynomorphs. Details of all of these, except the palynomorphs, are discussed in the relevant chapters herein. Of particular note is the mix of terrestrial, marine and freshwater forms (Storrs 1994), and the absence of groups that are abundant in the Rhaetian of southern Europe, such as cephalopods. The absence of cephalopods from the Penarth Group bone beds, as well as other typically marine invertebrates such as corals, suggests that the bone beds were probably not deposited under normal marine salinities. However, the few invertebrates that are recorded, such as bivalves, are certainly marine forms, albeit perhaps of a brackish nature, but these are rarely abundant.

Palaeoecological interpretation of the environment of deposition of the basal Westbury Formation bone bed using its contained fauna may be flawed, as the vertebrates are mostly derived from slightly older deposits (see below). Thus the vertebrates offer little concrete evidence to aid palaeosalinity interpretation. The presence of ichthyosaurs and plesiosaurs might be taken to suggest normal marine salinities but modern day aquatic reptiles such as crocodilians are capable of tolerating both freshwater and marine conditions. Indeed, the lack of very young plesiosaurs in marine black shales where adult plesiosaurs are abundant, suggests that perhaps plesiosaurs bred in fresh water. Likewise, the fish remains offer little evidence to support a marine origin. The sharks, whilst both diverse and abundant, could have been capable of tolerating both fresh and marine waters as do many species today. Perhaps of note is the extreme rarity of holocephalan dental plates in the bone beds (Duffin 1994). Holocephalans are exclusively marine and their rarity in the bone beds may be a reflection of reduced salinities. The dental plates of the dipnoan (lungfish) *Ceratodus* sp. (Text-fig. 5), although generally rare, are perhaps suggestive of fresh water, but their robust nature and their often worn surface (excluding *in vivo* wear) implies that they are derived, albeit probably not from a great distance. The presence of terrestrial vertebrates such as the dinosaurs and presumed freshwater aquatic reptiles such as the choristoderes (Storrs 1992) strongly suggests an input of material of fluvial origin. It is unknown whether the remains were brought in to the basin as dead carcasses or whether a bone bed of fluvial origin was reworked and incorporated with reworked material of marine or quasi-marine origin.

In addition, shark teeth attributable to the Carboniferous genera *Psammodus* and *Psephodus* were reported by Davis (1881), who considered that these genera may have survived until the Late Triassic. Note that Duffin reports here (Chapter 14) a number of derived Carboniferous taxa.

Among the invertebrates, Gould (1857) and Duffin (1978a) described remains of crustaceans occurring within the coprolites. The specimen described by Duffin (1978a) is fully articulated, an unusual condition for bone bed fossils and this must be attributed to its unusual occurrence within a coprolite. Sykes (1977) figured a number of bone beds of the Westbury Formation in which disarticulated, small bivalves form a significant component of the rock. These are most probably examples of *Isocyprina* (*Eotrapezium*) sp. and *Protocardia* sp.

VERTEBRATE TAPHONOMY

All of the vertebrate remains in the Westbury Formation bone beds are disarticulated, although some possible associations do occur. Teeth may sometimes occur within fragments of jaw and small accumulations of reptilian bones occur with no duplication of elements, suggesting derivation from one animal. Most skeletal elements show some degree of fracturing or abrasion but many teeth are well preserved. Some wear on shark teeth is *in vivo* rather than post-mortem. Most elements, including bones, have a polished surface. When fresh, most elements are shiny black or very dark brown but weather to a pale brown. A single unweathered slab may contain both black and brown bones.

There are often anomalous distributions of elements from certain taxa. For example, bones of the postcranial parts of the skeletons of most fishes are very rare, whereas bones that bear teeth are occasionally abundant. Teeth of the fish *Lepidotes* are known from the bone beds, but the scales of *Lepidotes* have yet to be identified with certainty. This is surprising since the scales of *Lepidotes* are often common in younger bone beds. It is not known why they should be absent from Penarth Group bone beds and it is possible that the teeth have been misidentified and that *Lepidotes* is not present in the Penarth Group ichthyofauna.

Coprolites are the most abundant phosphatic component of the bone beds, but the most common skeletal elements are teeth of sharks, bony fish and the enamel coated scales of bony fish. Bones and teeth of marine or aquatic reptiles are common but terrestrial reptiles are very rare.

DIAGENESIS

Macquaker (1994) discussed the diagenesis of Westbury Formation bone beds within the context of an overall study of the bone bed genesis. He noted the presence of a variety of diagenetic minerals, including quartz, pyrite, apatite and calcite. Each of these minerals is considered here with some general comments on their distribution and genesis. Some bone beds are unconsolidated and therefore difficult to observe in thin section and their diagenetic history may be distinct from that of the more easily observed (in thin section) indurated bone beds. Most of the comments here are based on observations of the bone bed at Aust Cliff, which may be more than 200 mm thick and is extremely hard.

Two cements are commonly found binding the clasts; pervasive calcite and patchily distributed pyrite (Pl. 8, figs 1–6). Calcite occurs as neo-morphosed microspar in void spaces within bones, on the surface of most clasts and as a neomorphic replacement of aragonitic bivalve shells (Pl. 8, fig. 4). There are some poikilotopic textures. Pyrite occurs as euhedral outgrowths from the bone bed, and pyritohedra may reach several mm in diameter (Text-fig. 7). Within the bone bed the euhedral pyrite is a pervasive cement (Pl. 8, fig. 6) and also replaces bivalves (Sykes 1977). Small pyrite crystals (<0·5 mm) often occur outlining neomorphosed bivalves.

Authigenic, bipyramidal quartz crystals have been reported from the bone bed by Antia and Sykes (1979). These euhedral grains are diagenetic overgrowths of silica on to formerly well-rounded quartz grains. Antia and Sykes (1979) supposed that the diagenetic quartz grew during diagenesis of the bone bed, since its distribution within the bone bed appears to be limited to the upper parts. However, it is worth noting that very many quartz grains lack any diagenetic overgrowths of silica and it is perhaps as likely that the bipyramidal grains obtained their overgrowths prior to incorporation in the bone beds.

Macquaker (1994) discussed the distribution of phosphate in the

TEXT-FIG. 7. Bone bed from the foreshore at Aust Cliff, Gloucestershire, with the fin spine of a shark and a rib bone in a groundmass of bone sand and grit cemented by calcite. Large masses of euhedral pyrite occur on the surface; University of Leicester, Department of Geology specimen LEIUG 19264; *c.* ×0·5.

Westbury Formation bone beds. He considered phosphate to occur as two distinct phases: (1) as originally phosphatic bioclasts such as bones and teeth; and (2) as phosphatic cements. However, I have not observed authigenic phosphate as a commonly occurring mineral cementing the clastic particles and it is possible that the phosphate cements as interpreted by Macquaker are in fact phosphatized coprolites and should be considered to be bioclasts. C. Duffin (pers. comm.) recorded authigenic collophane in a bone bed at Lavernock, and it is possible that its distribution is limited both stratigraphically and geographically.

Phosphatization of coprolites can be a result of defecation of material with high P levels into a neutral or slightly alkaline environment. This would result in rapid precipitation of any phosphate in solution within the faecal material, thus binding it and making it partially resistant to bacterial decay and macroscavenging. There is ample evidence that the phosphatization of the coprolites is a pre-bone bed event. Examples of weathered and broken coprolites are abundant within the bone bed at Aust Cliff (Text-fig. 6). In this section, such examples show brittle fracture, as well as fractured edges that lack weathering (Text-fig. 6). For brittle fracture to occur the coprolite must have been solid prior to incorporation in the bone bed, and for the fractured surface not to be weathered, any weathering must have predated the fracturing. It seems likely that no phosphate cements were precipitated within the bone beds and the model of Macquaker (1994) to explain such precipitation may not be applicable to the Westbury Formation.

C. Duffin (pers. comm.) reported the occurrence of authigenic glauconite within the pulp cavities of teeth from bone beds at several localities, but found no allogenic glauconite in the surrounding sediment. Thus the presence of glauconite restricted to cavity fills in bioclasts may be further evidence for reworking.

MODELS OF BONE BED GENESIS

Several models have been proposed for the genesis of bone beds and in particular for those of the Westbury Formation. Here the models are briefly reviewed and factors for and against their validity for the genesis of the Westbury Formation bone beds are considered. Any model for the genesis must account for all of the sedimentological, stratigraphical and biological components of the beds in the context of the Rhaetian transgressive event(s).

Mass mortality model

In this model, first proposed by Buckland (1829), large numbers of the biota must be killed at more-or-less the same time to provide quantities of bone material over and above that of the background mortality rate. Buckland (1829, p. 227) believed that sudden influxes of sediment were responsible for mass deaths, but Jukes-Browne (1912, p. 456) believed that repeated mass mortalities of both freshwater and marine vertebrates were caused in the early stages of the Rhaetian marine transgression when barriers between freshwater lakes and the sea were low and frequently transgressed by high tides. Lloyd-Morgan and Reynolds (1908) proposed a similar mass mortality model.

There is no taphonomic evidence to suggest that mass mortalities were important contributory processes for biogenic phosphate input to the Westbury Formation bone beds. However, due to the degree of reworking that occurred, the possibility of mass mortalities within a pre-Westbury Formation deposit from which the bioclasts were derived cannot be ruled out. Such a model must also account for the taxonomic variability and mixing of terrestrial, marine and fluvial faunas.

Winnowing model
This is a simplistic model, which requires episodes of erosion of the sea floor to rework sparsely distributed biogenic material causing it to be condensed to relatively high levels by density sorting. Several workers have proposed this as a model for the Westbury Formation bone beds (see above). It is a superficially attractive model that could account for many of the taphonomic features of the higher parts of the basal bone beds. It does not, however, explain the presence of terrestrial fauna nor does it account for the occurrence of poorly sorted bone beds. It may well be the main cause of genesis for those bone beds rich in small bivalves described by Sykes (1977).

Condensation model
This model requires that sedimentation rates were reduced to such a low level that the input of biogenic phosphate, largely in the form of bones and teeth, was sufficiently high over a period long enough to cover the sea floor with skeletal debris (Richardson 1901 and Wickes 1904 proposed a very high abundance of fish in rich feeding grounds). This model has the merit of explaining the high abundances of the marine vertebrate components but again it does not adequately explain the input of terrestrial forms or the lack of other elements of the fauna. Reduced rates of sedimentation would have allowed the bones and other clasts to have become bored or encrusted with epifauna. There is no evidence for this in any of the Westbury Formation bone beds.

Diagenetic concentration: the chemical lag model
In this model accumulations of skeletal debris are further concentrated by the progressive dissolution of more soluble material, notably aragonitic and calcitic faunal elements. Although this model cannot explain all of the features observed in the bone beds, it is clear that some bone beds contain a high abundance of shelly material, whereas others do not. For the most porous bone beds it is quite possible that large volumes of aragonitic shell material have been removed during burial diagenesis. Fresh samples of calcite-cemented bone bed from Lilstock show aragonitic bivalves preserved as external moulds, attesting to local selective dissolution. This model does not account for the faunal mixing and fragmentation.

Transgressive lag model
In this model, a transgressing sea derives clasts, including skeletal phosphates, from pre-existing sediments and incorporates them into first a basal bone bed and, further offshore from the site of erosion, into thinner intraformational bone beds. For the Westbury Formation, this material could have been derived from the top of the Blue Anchor Formation and as

a consequence, the vertebrate assemblage found within the Westbury Formation may not necessarily be representative of the Westbury biota. Indeed, this model could explain why ichthyosaurs and plesiosaurs are virtually unknown within the main mass of the Westbury Formation. However, the dominantly non-marine Blue Anchor Formation also does not yield ichthyosaur and plesiosaur bones and it is tempting to speculate that the Rhaetian transgression has removed completely a vertebrate bearing marine sequence of which there is no longer any *in situ* representative. This model was proposed by Macquaker (1994) and is the one that has most merit in explaining the stratigraphical distribution, faunal content, sedimentology and taphonomy of the Westbury bone beds. However, it is a composite model and draws on elements of many of the models discussed above.

Although it must remain speculation, a deposit perhaps faunistically similar to the Triassic Grenzebitumenzone of Ticino, Switzerland (e.g. see Rieppel 1985), which is famous for its abundant and exceptionally well preserved marine reptiles and fishes, may have been removed by erosion during the onset of the Late Triassic marine transgression. Such a hypothetical deposit could have been only partially lithified, was probably an organic-rich mudrock and probably lacked benthos. It may have been similar to many of the basal transgressive vertebrate-bearing black shales that are a prominent feature of marine transgressions in the overlying Jurassic.

DISCUSSION

Any model for bone bed genesis must explain the following:

1. The stratigraphical restriction

The stratigraphically narrow band in which bone beds are developed implies that the combination of factors responsible for bone bed genesis could only be achieved over a narrow time frame. As the bone beds occur in a transgressive sequence, it is tempting to speculate that the presence of bone beds in the Westbury Formation is at least in part depth controlled. Water depth would have controlled the amount and degree of reworking that could occur due to storm and wave activity. Furthermore, concomitant reworking of pre-existing coastal plain sequences during a transgression would take place along the entire transgressing front. If part of the pre-existing coastal plain sequence contained vertebrate-bearing horizons, the more resistant elements of these would become incorporated into the earliest sediments of the transgression event.

2. The wide geographical distribution

The wide geographical distribution of bone beds towards the base of the

Westbury Formation (they occur intermittently along a linear outcrop of more than 400 km) implies that the conditions responsible were widespread and presumably relatively uniform. It is unlikely that currents would have been involved as water currents operating over such long distances would have produced more sorting than is seen. Storm activity (over a very wide area in the form of large hurricanes) could have been responsible for causing concentrations of biogenic materials without causing a lot of sorting (Short 1904).

3. The sedimentological characteristics, including geochemical characters
Sedimentological characteristics are often distinct for the different bone beds, but a number of features are common to all. Some dissimilarities are due to later diagenetic processes, especially the degree of cementation. An important similarity is that all of the bone beds are coarser grained than the surrounding or overlying sediments. In addition, all contain vertebrate remains, usually in greater abundance than invertebrates. Differences are usually in the overall grain size and as such may reflect localized variation in hydraulic energy or proximity to palaeoshorelines. It is notable that many of the bone beds high in the Westbury Formation lack the larger reptile remains, and are dominated by small teeth and scales of fishes. Some of the basal bone beds are conglomeratic but the largest clasts are usually ripped up from the underlying formation. In most cases the lower boundary of the bone beds is sharp but the upper boundary is often gradational.

4. The faunal and floral characteristics
The fauna of the bone beds is distinct from that of the surrounding shales. Vertebrates are rare in the shales of the Westbury Formation except within the bone bed horizons. It would not be possible to rework the shales to generate the abundances found in the bone beds. A significant proportion of the bone bed taxa are non-marine and of either freshwater or terrestrial origin. Thus, a component of the bone beds must have been introduced as either drifting carcasses from nearby river systems or by reworking from pre-existing vertebrate bearing sequences that were destroyed or remain unidentified. If drifting carcasses introduced material into the basin, it seems odd that no isolated occurrences of better articulated skeletons have been discovered in the Westbury Formation shales. Furthermore, the general lack of larger skeletal elements in those bone beds above the basal bone bed suggests that drifting of carcasses was not a major contributing factor for introducing terrestrial and fluvial skeletal components into the basin. The scarcity of macrophyte material (often abundant in vertebrate-bearing accumulations) is perplexing, but may be a taphonomic filtering effect (see below).

5. The taphonomic characteristics

The most notable taphonomic effects observed in the bone beds are: (1) the lack of articulated remains; (2) the mix of terrestrial and marine components; (3) the relative abundance of hard resistant components such as shark teeth and enameloid scales; (4) the absence of certain skeletal elements for some taxa known to have preservable elements; (5) the presence of pristine bone adjacent to highly worn bone; and (6) the high abundance of coprolites.

Some of these observations are easy to explain. The high abundance of coprolites is almost certainly an effect of natural abundance. The abundance of teeth is due to the high number of teeth produced by sharks as well as their resistance to abrasion. Teeth may also be less susceptible to bioerosion than bone. Observations that are more difficult to explain are the presence of worn and unworn bones in the same deposit. If both elements were introduced into the bed at the same time from a dead animal, then it might be expected that they would be worn to a similar degree. It is possible that the most worn elements were the first to be introduced into the system and that the least worn elements were the 'last in'. However, this simplistic scenario is highly unlikely, requiring evidence of mixing with beds that show little or no sorting. In any environment where physical transport was prolonged, a high degree of sorting of material could be expected and this is certainly not the case. It is much more likely that these elements are derived and those that are worn have been derived from different horizons from those that are unworn. The absence of certain skeletal elements and the lack of articulated components can also be explained by derivation from pre-existing deposits.

6. The diagenetic effects

These operate to a lesser degree, as they postdate deposition.

CONCLUSIONS

The bone beds of the Westbury Formation, although locally discontinuous, are present over a very wide area of the outcrop and occur consistently as either basal bone beds, or within the lowest few metres of the formation. They comprise an abundance of derived bioclasts consisting mainly of teeth and scales of fish, abundant coprolites and isolated bones of marine and terrestrial reptiles. The vertebrate assemblage is diverse and comprises a wide variety of terrestrial, freshwater and marine forms, including elasmobranchs, actinopterygians, dipnoans, ichthyopterygians, sauropterygians, chorsitoderes and archosaurs. Invertebrates are generally rare but occur as low diversity assemblages of disarticulated bivalves and gastropods and more rarely as articulated inclusions within coprolites.

6. FORAMINIFERA

by DAVID M. MARTILL *and* ANDREW SWIFT

Foraminifera are single-celled protozoa which produce protective tests either by surrounding themselves with tiny grains of sediment cemented together (agglutinated or 'arenaceous' forms), or by secreting what is usually a calcitic (rarely aragonitic) chambered test, often of elaborate design. These are the remains which are found as fossils. Today, both benthic and planktic species are found world-wide, but most Penarth Group forms were probably benthic. Penarth Group specimens occur in both limestones and clastic rocks and fall into the 0·2–1·0 mm size range. As with ostracods and other microfauna, extraction of foraminifera from softer clastic rocks may be accomplished by simple boiling in water, followed by sieving. For more advanced extraction procedures see Brasier (1980). Foraminifera are best studied with a microscope, but large specimens can be seen with a hand lens and exceptionally may be detectable with the naked eye.

Foraminifera are characteristic of marine environments and are important both for palaeoenvironmental analysis and as index fossils in biostratigraphy. However, most Penarth Group foraminifera belong to species which range into the Lower Jurassic and may thus be considered to be of little value for fine-scale biostratigraphy. Nevertheless, their first appearances have proved to be useful, as different species appear at progressive intervals through the group, with such occurrences being noted over a wide geographical area. Thus, foraminifera are of some value for correlation.

Despite their common and widespread occurrence, especially in the higher beds of the Penarth Group, foraminifera have received little attention. They were first reported from the 'Rhaetic' by Chapman (1894), but earlier, Strickland (1846) described two species from the 'Lias', one of which was *Polymorphina* (*=Eoguttulina*) *liassica*, a very common Penarth Group form. Chapman (1894) described a diverse fauna dominated by agglutinating forms from clays and a sandstone associated with the 'Wedmore Stone' facies of the lower Westbury Formation at Wedmore, Somerset. He also noted foraminifera in thin sections of the 'Wedmore Stone' itself. No less than 26 species of nine genera were described, many of which were attributed to Carboniferous taxa. Nine species of the genus *Ammodiscus* alone were recognized and there can be little doubt that many of Chapman's (1894) species are synonymous. Resolution of such taxonomic problems is constrained by the poor illustrations (drawings) and the apparent disappearance of Chapman's collections. After Chapman's

(1894) work, there was a long hiatus during which no studies of Penarth Group foraminifera were published. Barnard (1950) recorded only 'doubtful casts' of three genera from the '*Pteria contorta* shales' (i.e. Westbury Formation *sensu stricto*, but Barnard may have have meant the whole of the 'Rhaetic') and 'succeeding *Ostrea* and *planorbis* beds' (i.e. lowest Lias Group) of the Dorset coast. Edmunds *et al.* (1965) recorded *Dentalina pseudocommunis*, *Eoguttulina liassica* and *Lingulina lanceolata* from the Langport Member and *Eoguttulina liassica*, *Lenticulina* aff. *muensteri* and *Nodosaria* cf. *nitidana* from the very top of the Cotham Member in Warwickshire. A fossiliferous shale at a comparable horizon in the Cotham Member is found in many areas in the Midlands as far north as Bantycock Quarry near Newark in Nottinghamshire; it contains both foraminifera and ostracods, as well as macrofossils. The most foraminifera-rich horizons in the Langport Member are to be found in the siltstones and mudstones of the 'Watchet Beds' at Lavernock Point near Penarth; two particular horizons (a few tens of millimetres from the base and approximately 1 m from the top) yield rich faunas, together with ostracods of several genera (Ivimey-Cook 1974, pers. comm.).

Banner *et al.* (1971) discussed the usefulness of forms in the *Geinitzinita* plexus (Penarth Group forms in this group are now assigned to *Lingulina*) for recognizing Rhaetian as opposed to Hettangian levels around Barry in South Wales. A long overdue assessment of Penarth Group foraminifera was published by Copestake (1981), but it was not until the wide-ranging revision by the same author (Copestake 1989) that the many taxonomic and biostratigraphical problems were addressed. This study has provided much of the information in this section, supplemented with recent work by one of the authors (AS). An important part of the revisions of Copestake (1981, 1989) was a re-evaluation of Chapman's (1894) pioneering work. Initial doubts (Copestake 1981) as to the authenticity and provenance of Chapman's specimens which were referred to Upper Palaeozoic taxa were partly resolved later (Copestake 1989) and the association is now considered to be typical of the Westbury Formation in south-west Britain, notwithstanding doubts still attending certain forms assigned to Upper Palaeozoic genera such as *Stacheia* and *Nodosinella*.

In general terms the Westbury Formation is dominated by agglutinating forms, many of which range into the Cotham Member and higher. In the Cotham Member calcareous forms become increasingly important, perhaps in response to the greater availablility of calcium carbonate for test building. By the time of the deposition of the highest beds of the Cotham Member a diverse population of calcareous forms had become established, typified by abundant *Eoguttulina liassica*. Representatives of this assemblage continued to dominate Langport Member faunas, with the significant addition of zonally important forms such as members of the *Lingulina tenera* complex. Copestake (1989) erected a tripartite zonal scheme for the interval from the higher beds of the Blue Anchor Formation to the base of the Jurassic consisting of two assemblages, followed by a third zone which spans the Triassic-Jurassic boundary. At least 17 species in 11 genera were listed by Copestake (1989, figs 5·6–5·7), but it is unlikely that these are the only forms present in the Penarth Group. For instance, Hallam and El Shaarawy (1982) listed 13 species in eight genera from the Cotham Member and higher Langport Member ('Watchet Beds'), very few of which match up with those of Copestake (1989). Hallam and El Shaarawy (1982) did not figure or describe their collections so comparisons and possible synonymies cannot be evaluated, and they also reported that the greatest 'density' and species richness were to be found in the Cotham Member at Lavernock Point, a conclusion at odds with the findings of other researchers. Amongst the foraminifera recorded by Edmunds *et al.* (1965), *Lenticulina* aff. *muensteri* (Roemer) and *Lingulina lanceolata* (Haeusler) were mentioned by Copestake (1981) but not subsequently in his revision (Copestake 1989). Hallam and El Shaarawy (1982) also did not note these species. Thus, even allowing for different taxonomic interpretations, the number of foraminiferan taxa present in the Penarth Group is unknown and potentially large, leaving much scope for further work. Only the more common forms are figured in this review.

The suprageneric classification employed here follows Loeblich and Tappan (1984). All figured specimens are held in the A. Swift collection, Department of Geology, University of Leicester. For illustrations of the taxa not figured here the reader is referred to the original citations, Franke (1936) and Copestake (1989).

SPECIES DESCRIPTIONS

Order FORAMINIFERIDA von Eichwald, 1830
Suborder TEXTULARIINA Delage and Hèrouard, 1896
Superfamily ASTRORHIZOIDEA Brady, 1881
Family BATHYSYPHONIDAE Avnimelech, 1952
Genus BATHYSIPHON M. Sars, *in* G. O. Sars, 1872
Bathysiphon spp.

Description. Test agglutinated, highly elongate, slender. Recent species can reach 50 mm long. There may be annular constrictions. Test may be closed at one end.
Remarks. *Bathysiphon* ranges from the Cambrian to Recent and is cosmopolitan. In the Penarth Group there are local acmes of species of *Bathysiphon* at the base of the Cotham Member, and in the Langport Member around the junction of the 'White Lias' and 'Watchet Beds' (Copestake 1989).

Superfamily AMMODISCOIDEA Reuss, 1862
Family AMMODISCIDAE Reuss, 1862
Genus AMMODISCUS Reuss, 1861
Ammodiscus auriculus (Chapman, 1894)
Plate 9, figure 1

Description. Test agglutinated, consisting of up to three planispiral, slightly ovate whorls. Test thin-walled, usually appears white.

Remarks. It is possible that several species of *Ammodiscus* and related genera are present in the Penarth Group, as such forms occur throughout the marine part of the sequence and are common as pyritic internal moulds in acid-resistant residues. Chapman (1894) described several species of *Ammodiscus*, but Copestake figured only the ovate-whorled *A. auriculus* (Pl. 9, fig. 1) and implied the synonymy of all or most of Chapman's species. An example of a circular-whorled ammodiscid is illustrated on Plate 9, figure 2, but specific assignment is not possible.

Distribution. Westbury Formation and Cotham Member at Watchet, Westbury Formation at Wedmore. At Lavernock it has been found in the Langport Member and ranges into the Pre-planorbis Beds.

Genus GLOMOSPIRA Rzehak, 1888
Glomospira perplexa Franke, 1936

Description. Test is an agglutinated, irregularly coiled tube, in which whorls are in contact. Overall morphology sub-globular.

Distribution. Westbury Formation and Cotham Member at Watchet and St Audries Slip. Ranges into the Pleinsbachian (Lower Jurassic).

Glomospira subparvula Bartenstein, 1962

Description. Test agglutinated, with a coiling habit like that of miliolid foraminifera, i.e. somewhat elongate, spindle shaped and coiled in three planes.

Distribution. Common in the Westbury Formation at Watchet and South Glamorgan. Also known from the Cotham Member.

Genus GLOMOSPIRELLA Plummer, 1945
Glomospirella sp. 1

Description. Test agglutinated, ovate, spiral, with early whorls globose and last whorls more or less planispiral.

Remarks. Can easily be confused with the similar *Ammodiscus auriculus*. Note that *Glomospirella* is not truly planispiral.

Distribution. Westbury Formation and Cotham Member.

Family HORMOSINIDAE Haeckel, 1894
Subfamily REOPHACINAE Cushman, 1910
Genus REOPHAX de Montfort, 1808
Reophax helvetica (Haeusler, 1881)

Description. Fine- to medium-grained agglutinated test with five uniserial, overlapping chambers. Sutures between overlapping chambers straight. Late chambers equidimensional, but early chambers broader than high.
Distribution. Westbury Formation at Lavernock and Langport Member at Watchet.

Family LITUOLIDAE de Blainville, 1827
Genus AMMOBACULITES Cushman, 1910
Ammobaculites cf. *eiseli* Paalzow, 1935

Description. Test coarsely agglutinated. Overall form elongate, rectilinear, with divergent chambers.
Remarks. Although *A.* cf. *eiseli* has only been recorded from the Cotham Member of the Penarth Group, the species *A. eiseli* is recorded from the Upper Permian elsewhere in Britain (Pattison 1981). Thus, this foraminiferan is perhaps one of the few species to pass through the Permian-Triassic boundary in the UK. Another indeterminate species of *Ammobaculites* was recorded by Copestake (1989) from the Somerset coast.

Superfamily TROCHAMMINOIDEA Schwager, 1877
Family TROCHAMMINIDAE Schwager, 1877
Subfamily TROCHOMININAE Schwager, 1877
Genus TROCHAMMINA Parker and Jones, 1859
Trochammina squamosa Ziegler, 1964

Description. Test coarsely agglutinated, consisting of three whorls. Final whorl with between four and six large, broad chambers.
Distribution. Copestake (1989) recorded this form from between the top of the Westbury Formation and the uppermost part of the Langport Member at Watchet, and throughout the section from the uppermost part of the Westbury Formation to the basal Lias Group at Lavernock.

Suborder LAGENINA Delage and Hèrouard, 1896
Superfamily NODOSARIOIDEA Ehrenberg, 1838
Family NODOSARIIDAE Ehrenberg, 1838
Subfamily NODOSARIINAE Ehrenberg, 1838
Genus DENTALINA d'Orbigny, 1826
Dentalina pseudocommunis Franke, 1936
Plate 9, figures 3–5

Description. Test hyaline, uniserial, narrow and elongate; may be arcuate or almost straight. Early chambers broad but later chambers become high. Aperture radiate, sutures oblique.

Remarks. One of a group of arcuate nodosariids which occur commonly, sometimes in large numbers at certain levels, especially higher in the Penarth Group. The number of species has never been delineated, and the illustrated forms do not compare with *D. pseudocommunis sensu stricto*, lacking the oblique sutures and flattened inside edge to the chambers (see *D. ventricosa* Franke, 1936, p. 33, pl. 3, fig. 1 for a similar form).

Distribution. *D. pseudocommunis* is a long-ranging species, appearing in the Blue Anchor Formation and continuing throughout the Penarth Group into the Lower Jurassic.

Genus NODOSARIA Lamarck, 1812
Nodosaria sp.

Description. Nodosariids are generally elongate and rectilinear with globular, strongly ribbed chambers.

Remarks. Copestake (1989) recorded an unnamed species of *Nodosaria* from the Langport Member, but the first named species referred to the genus, *N. metensis*, is not known from lower than the late Rhaetian Preplanorbis Beds. However, Ivimey-Cook (pers. comm.) stated that forms

---

EXPLANATION OF PLATE 9

Fig. 1. *Ammodiscus auriculus* Chapman, 1894; AS 32/35; Langport Member; Gotham Hills, Nottinghamshire.

Fig. 2. *Ammodiscus* sp.; AS 27/8; top of Langport Member; Bantycock Quarry, Newark, Nottinghamshire.

Figs 3–5. *Dentalina* aff. *pseudocommunis* Franke, 1936. 3, AS 22/10; top of Cotham Member; Bantycock Quarry, Nottinghamshire. 4, AS 6219/26; Langport Member; Gotham Hills, Nottinghamshire. 5, AS 22/9; top of Cotham Member; Bantycock Quarry, Nottinghamshire.

Figs 6–7. *Lingulina cernua* (Berthelin, 1879); top of Cotham Member; Bantycock Quarry, Nottinghamshire. 6, AS 22/8. 7, AS 27/4.

Figs 8–9. *Lingulina tenera* subsp.; top of Cotham Member; Long Itchington Quarry, Warwickshire. 8, AS 29/11. 9, AS 21/17.

Figs 10–12. *Eoguttulina liassica* (Strickland, 1846). 10, AS 22/15; top of Langport Member; Bantycock Quarry, Nottinghamshire. 11, AS 22/16; top of Langport Member; Bantycock Quarry, Nottinghamshire. 12, AS 6219/22; Langport Member; Gotham Hills, Nottinghamshire.

Figs 13–14. *Incertae sedis*; lower Langport Member; Culverhole Point, between Seaton and Lyme Regis, Devon. 13, AS 29/4. 14, AS 23/24.

All ×100.

PLATE 9

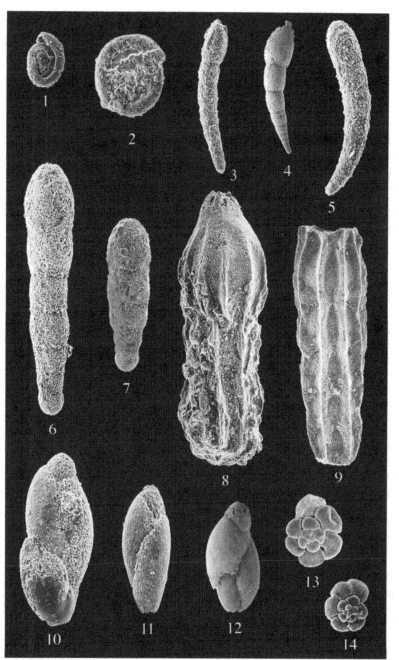

very similar to *N. columnaris* Franke (1936, p. 48, pl. 4, fig. 19) occur in the Langport Member at Lavernock Point.

Subfamily LINGULININAE Loeblich and Tappan, 1961
Genus LINGULINA d'Orbigny, 1826
Lingulina cernua (Berthelin, 1879)
Plate 9, figures 6–7

Description. Test elongate, slightly curved, smooth, slightly compressed to circular in cross section. Sutures between chambers depressed, either straight or slightly curved. Length variable.
Remarks. Penarth Group specimens tend to be almost straight and of variable length. The two illustrated examples show this variation.
Distribution. A familiar form in the Lower Jurassic, this species first appears in Britain towards the top of the Cotham Member.

Lingulina tenera tenera Barnard, 1956

Description. Test elongate, slightly compressed with marginal keel and two prominent longitudinal ribs on each side. Terminal aperture slit-like.
Remarks. One of the earliest of the lingulinid foraminifera. There are a number of subspecies and varieties of *L. tenera*, and recognition requires experience of the group. *L. tenera tenera* is distinguished by two strong ribs on either side of the test, which lack intervening striations. Two members of the *L. tenera* plexus are illustrated on Plate 9, figures 8–9, and despite being incomplete, are not apparently comparable to any of the subspecies recorded by Copestake (1989) or those described here. The specimens were recovered from the top of the Cotham Member at Long Itchington Quarry, which makes them the earliest members of the plexus in the Penarth Group.
Distribution. According to Copestake (1989) *L. tenera tenera* first occurs in the higher beds of Langport Member ('Watchet Beds') and ranges into the Toarcian (Lower Jurassic).

Lingulina tenera collenoti (Terquem, 1866)

Description. Similar to *L. tenera tenera*, but with ribs of equal prominence and test margin rounded.
Remarks. A 'variety' of *L. tenera* known as *L. tenera* var. *octocostata* is known from the Langport Member in Somerset and at Lavernock Point. It is recognizable by having two pairs of strong ribs along the test. Copestake (1989) also recorded *Lingulina* cf. *tenera* from the Devon coast.

Family POLYMORPHINIDAE d'Orbigny, *in* de la Sagra, 1839
Subfamily POLYMORPHININAE d'Orbigny, *in* de la Sagra, 1839
Genus EOGUTTULINA Cushman and Ozawa, 1930
Eoguttulina liassica (Strickland, 1846)
Plate 9, figures 10–12

Description. Test elongate, smooth with merging chambers arranged spirally. Aperture terminal, central, radiate.

Remarks. A highly variable foraminiferan which may be biserial. Some of the more divergent polymorphs of this species have been given different species names. Copestake (1989) preferred to unite all polymorphs in one taxon as the variation is continuous. *E. liassica* can be very common at certain levels, particularly in both units of the Lilstock Formation.

Distribution. Occurs throughout the Penarth Group and ranges upwards into the Lower Jurassic.

Suborder ROBERTININA Loeblich and Tappan, 1984
Family CERATOBULIMINIDAE Cushman, 1927
Subfamily REINHOLDELLINAE Seiglie and Bermúdez, 1965
Genus REINHOLDELLA Brotzen, 1948?
Reinholdella? planiconvexa (Fuchs, 1970)

Description. Test smooth, discoidal to ovate, plano-convex in vertical section. Marked umbilical hollow. Clear supplementary aperture on ventral side. Two to two-and-a-half whorls, final whorl with five or six chambers (Copestake and Johnson 1981).

Remarks. Copestake (1989) tentatively placed foraminifera similar to *R. planiconvexa* in this taxon because the British specimens are the first to be reported from rocks older than the Lower Jurassic.

Distribution. First occurs at the base of the Langport Member and ranges into the Lower Jurassic.

Incertae sedis
Plate 9, figures 13–14

Remarks. Pyritized internal moulds of a species of a very small, calcareous, low trochospiral foraminiferan are relatively common in acid resistant residues of Langport Member limestones, and two examples are illustrated (Pl. 9, figs 13–14). It is not possible to identify such forms, since the taxonomy of the taxon cannot be established from moulds only, but the specimens are otherwise well-preserved. Such general morphology is most similar to that of members of the superfamily Discorboidea of the suborder Rotaliina. Barnard (1950, p. 377, fig. 9) illustrated a taxon that he named *Epistomina liassica* from the Lias Group of the Dorset coast, which bears some resemblance to this Penarth Group form. Warrington (1978, pl. 3, fig. 40) also figured a closely similar form from the Westbury Formation of the Withycombe Farm Borehole, near Banbury.

OTHER TAXA

The following taxa, not in Copestake (1989), were recorded from the Cotham Member and higher Langport Member ('Watchet Beds') by Hallam and El Shaarawy (1982), but their validity and possible synonymies have not been confirmed:

Astacolus terquemi (d'Orbigny) (?=*Dentalina terquemi* d'Orbigny, also listed by Hallam and El Shaarawy 1982)
Dentalina curva Liebus
Eoguttulina bilocularis (Terquem)
E. kuhni Franke
Frondicularia brizaeformis Bornemann
F. dentaliformis Terquem
Lingulina laevissima (Terquem)
Nodosaria nitidana Brand (a form named *Nodosaria* cf. *nitidana* was listed by Edmunds *et al.* 1965)
Paleopolymorphina vagina (Terquem)
Vaginulina constricta (Terquem and Berthelin)

7. GASTROPODS

by MICHAEL J. BARKER and MARTIN C. MUNT

The gastropods of the Penarth Group all fall into the small and micromorphic size range. They are sporadically distributed or rare, with published descriptions frequently based upon single specimens which are often incomplete. These factors have reduced their appeal for further systematic re-examination, which is long overdue.

The Triassic was an important time in gastropod evolution, with the emergence of many new higher taxa. Amongst these, the Neritoidea are represented in the Penarth Group and indeed *Naticopsina oppelii* can be locally abundant (see remarks below).

Early works on the gastropod fauna of the Penarth Group are few, but two references are significant. All the taxa described and figured in this work were first described and illustrated by Moore (1861*b*) from his 'Flinty bed' at Beer Crowcombe (=Beer Crocombe of the present day) in Somerset. This bed is a sparry bioclastic limestone thought to be located at a horizon low in the Westbury Formation. As with the bivalves (see Chapter 8), it should be noted that the fossil material attributed to the 'Flinty bed' was obtained from mixed spoil heaps from a canal tunnel excavation and therefore some doubt must always attend the actual horizon or horizons from which the specimens originated. In mitigation, the matrix surrounding the specimens is generally consistent and specimens of the characteristic Westbury Formation bivalve *Rhaetavicula contorta* were also recovered. Some revisions were proposed by Hudleston and Wilson (1892) when they listed Moore's taxa with their synonymies.

The primary taxonomic nomenclature used in this work originates from these sources with appropriate revisions based largely on the numerous works of Cossmann, especially the *Essais de Paléoconchologie Comparée* (13 volumes, 1895–1925). Higher taxonomic nomenclature follows that of Tracey *et al.* (1993) and/or Ponder and Warén (1988), with morphological descriptive terms after Cox (1960).

The specimens on which the following descriptions are based are housed in the Charles Moore collection at the Bath Royal Literary and Scientific Institute (BRLSI M 37–M 39, M 41–M 43, M 45–M 52), as are all the figured specimens. Elsewhere, important collections of Penarth Group gastropods are housed in The Natural History Museum, London, the British Geological Survey at Keyworth near Nottingham and the Bristol City Museum and Art Gallery.

SPECIES DESCRIPTIONS

Subclass STREPTONEURA Spengel, 1881
Order EUOMPHALINA de Koninck, 1881
Superfamily EUOMPHALOIDEA de Koninck, 1881
Family EUOMPHALIDAE de Koninck, 1881
Genus DISCOHELIX Dunker, 1847
Discohelix suessii (Moore, 1861*b*)
Plate 10, figure 1

Material. BRLSI M 37, the holotype and only known specimen, is a poorly preserved recrystallized shell which has been broken and repaired. The umbilicus is missing and only the outer whorl is preserved reasonably intact. It is now glued onto card so that only one side is visible. See Moore (1861*b*, p. 511) for the original description prior to breakage and mounting.
Description. Small discoidal shell up to 10 mm in diameter and 4 mm high. Abapical surface reportedly concave (see above). Tops and bottoms of whorls flattened and sides strongly convex. Suture impressed. Surface of last whorl has 12 visible axial nodes/tubercles (projected 15 or 16 per whorl), which flatten adaxially. These nodes/tubercles situated more on periphery of whorl than in Moore's figure (1861*b*, pl. 15, fig. 3) and also extend to inside of the whorl. Rounded whorl sides carry three equidistant strong spiral threads but these are undulose rather than straight. Faint spiral ornament visible on outer whorl. Aperture not visible.
Distribution. Westbury Formation, Beer Crocombe, Somerset.

Order ARCHAEOGASTROPODA Thiele, 1925
Suborder VETIGASTROPODA Salvini-Plawén, 1980
Superfamily TROCHOIDEA Rafinesque, 1815
Family TROCHIDAE Rafinesque, 1815
Genus SOLARIOCONULUS Cossmann, 1918
Solarioconulus waltonii (Moore, 1861*b*)
(=*Solarioconulus nudus* (Moore, 1861*b*) (*non* Münster, 1833))
Plate 10, figure 2

Material. BRLSI M 38, the holotype of *Trochus nudus* Moore (*non* Münster), has the apex missing with three-and-a-half whorls visible. BRLSI M 39, the holotype of *Trochus waltonii* Moore, is better preserved than BRLSI M 38 and comprises three-and-a-half whorls.
Description. Small trochiform/conical shell up to 6 mm high. Up to five shallowly convex whorls ornamented with strong prosocyrt striae at *c.* 45° from the suture. Five fine spiral striae on base of shell. Some faint spiral ornament visible on adapical and abapical sides of whorls. Suture impressed and grooved on adapical side. Angular periphery of spire and base slightly rounded.

Remarks. This species was originally referred to the genus *Trochus*, but probably belongs to *Solarioconulus* Cossmann, 1918. The two species (*S. nudus* and *S. waltonii*) are undoubtedly synonymous, as suggested by Dittmar (1864, p. 146) and Hudleston and Wilson (1892, p. 139). The differences are wholly attributable to the state of wear, BRLSI M 39 being much less worn. The drawing of *S. waltonii* in Moore (1861*b*, pl. 16, fig. 23) is poor; the whorls are not as convex as depicted. *S. waltonii* is retained by page priority.

Distribution. Westbury Formation, Beer Crocombe, Somerset. Westbury Formation, Collin Glen, Antrim, near Belfast (Ivimey-Cook 1975).

Suborder NERITIMORPHA Golikov and Starobogatov, 1975
Superfamily NERITOIDEA Rafinesque, 1815
Family NERITOPSIDAE Gray, 1847 (*fide* Moore, 1960)
Subfamily NATICOPSINAE Cossmann, 1896
Genus NATICOPSINA M'Coy, 1844
Naticopsina oppelii (Moore, 1861*b*)
Plate 10, figure 3

Description. The single cited specimen (BRLSI M 43, the holotype) is a small globular neritiform shell 5 mm high, but heights up to 8 mm are quite common. Whorls strongly convex, suture not impressed. Growth lines slightly prosocyrt, prominent. Shell smooth where outer shell layer lost, otherwise very faint spiral ornament. Aperture large, ovate. A diffuse central dark band on penultimate whorl may represent relict colour banding.

Remarks. This is the only Penarth Group neritiform gastropod. *N. pylensis* Tawney, 1866 is conspecific with *N. oppelii*. The drawing of *N. oppelii* in Moore (1861*b*, pl. 16, fig. 17) is reasonably accurate. The most abundant gastropod in the Penarth Group, often found crowded on bedding planes at certain levels in the Westbury Formation. Commonly occurs as fibrous calcite or pyrite films.

Distribution. Westbury Formation at Beer Crocombe and Watchet (Somerset), Collin Glen (near Belfast), Port More borehole, Larne borehole, Felixkirk borehole (Yorkshire). Cotham Member at Wells, Somerset and Pylle Hill, Bristol. Also occurs commonly at Pyle (near Bridgend) and at Penarth, South Wales.

Order HETEROSTROPHA Fischer, 1880–1887
Superfamily ARCHITECTONICOIDEA Gray, 1850
Family MATHILDIDAE Dall, 1889*b*
Genus PROMATHILDIA Andreae, 1887
Promathildia rhaetica (Moore, 1861*b*)

PLATE 10

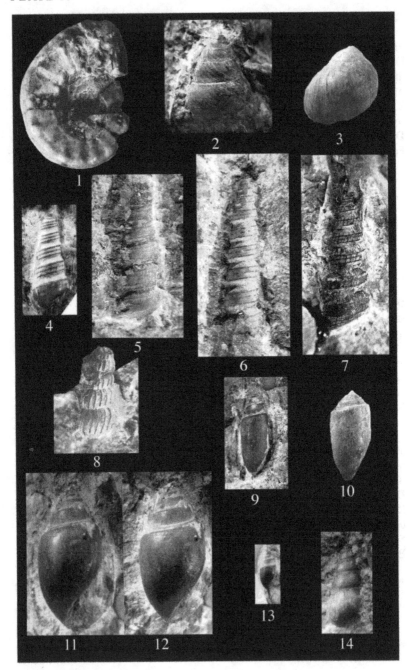

(=*Promathildia constricta* (Moore, 1861*b*) (*non Cerithium constrictum* Deshayes, 1831); =*Promathildia cylindrica* (Moore, 1861*b*))
Plate 10, figures 4–6

Material. BRLSI M 45, the holotype of *P. constricta*, is a single, rather worn specimen. The drawing of Moore (1861*b*, pl. 16, fig. 13) is accurate. BRLSI M 47, the holotype of *P. cylindrica*, has nine whorls visible. BRLSI M 48, the holotype of *P. rhaetica*, is the best preserved and therefore, in spite of page priority, is taken as the name of the taxon.

Description. Small, elongate, turreted shell 6 mm high, comprising up to nine whorls. Whorls convex with rounded appearance, suture impressed on broad ramps. Ornament comprises three equidistant spiral threads situated in a zone just below mid-whorl. Where the shell is well-preserved, the spiral threads are sharp crested and whorls are crossed by prominent prosocyrt growth lines. Aperture round.

Remarks. The distinctive ornament suggests that this rare species should probably be assigned to the genus *Promathildia* (see Cossmann 1912, p. 3). Moore's (1861*b*) three species are synonymized after examination of all his known material. BRLSI M 47, the holotype of *P. cylindrica*, has a slightly smaller apical angle but otherwise the ornamentation and shell proportions are identical in all. The best preserved specimens come from fresh levels in borehole cores.

Distribution. Westbury Formation at Beer Crocombe, Somerset and Pylle Hill, Bristol. Specimens recovered from the British Geological Survey

---

EXPLANATION OF PLATE 10

All from the Moore Collection from the 'Flinty Bed' (lower Westbury Formation) at Beer Crocombe, Somerset

Fig. 1 *Discohelix suessii* (Moore, 1861*b*); BRLSI M 37.

Fig. 2. *Solarioconulus waltonii* (Moore, 1861*b*); BRLSI M 38.

Fig. 3. *Naticopsina oppelii* (Moore, 1861*b*); BRLSI M 43.

Fig. 4. *Promathildia rhaetica* (Moore, 1861*b*) =*Promathildia constricta* (Moore, 1861*b*) (*non Cerithium constrictum* Deshayes, 1831); BRLSI M 45.

Fig. 5. *Promathildia rhaetica* (Moore, 1861*b*); BRLSI M 48.

Fig. 6. *Promathildia rhaetica* (Moore, 1861*b*) =*Promathildia cylindrica* (Moore, 1861*b*); BRLSI M 47.

Fig. 7. *Promathildia?* *decorata* (Moore, 1861*b*); BRLSI M 46.

Fig. 8. *Zygopleura henrici* Martin, 1860; BRLSI M 41.

Fig. 9. *Cylindrobullina elongata* (Moore, 1861*b*); BRLSI M 50.

Fig. 10. *Cylindrobullina oviformis* (Moore, 1861*b*); BRLSI M 51.

Figs 11–12. *Cylindrobullina ovalis* (Moore, 1861*b*); BRLSI M 49.

Fig. 13. *Cylindrobullina fusiformis* (Moore, 1861*b*); BRLSI M 52.

Fig. 14. '*Cheminitzia granum*' Dittmar, 1864; BRLSI M 42.

All ×4.

boreholes at Wilkesley and Platt Lane (Cheshire) are probably referable to this taxon.

Promathildia? *decorata* (Moore, 1861*b*)
Plate 10, figure 7

Material. BRLSI M 46, the holotype, is partially obscured by matrix. The drawing of Moore (1861*b*, pl. 16, fig. 14) is embellished as the full outline is not visible.
Description. Small, elongate, turreted shell about 9 mm high, comprising seven whorls. Whorls convex, suture impressed. Ornament comprises three spiral threads crossed by axial ornament. Axial ornament aperturally concave and same strength as the spiral threads, giving a reticulate ornament which is not nodose at intersections. Three spiral lines on base.
Remarks. This very rare species should probably be assigned to the genus *Promathildia*.
Distribution. Westbury Formation, Beer Crocombe, Somerset.

Order APOGASTROPODA Salvini-Plawén and Haszprunar, 1987
Suborder DISCOPODA Fischer, 1880–87
Superfamily LOXONEMATOIDEA Koken, 1889 (n. t. Cossmann, 1909)
Family ZYGOPLEURIDAE Wenz, 1938–44 (as Zygopleurinae, n. t. Cox 1960)
Genus ZYGOPLEURA Koken, 1892
Zygopleura henrici Martin, 1860
Plate 10, figure 8

Description. Small turreted shell up to 6 mm high, comprising five visible whorls. Whorls moderately convex, suture impressed. Distinctive ornament comprises prominent, slightly curved (lunulate/falcate) axial ribs.
Remarks. The drawing of *Z. henrici* in Moore (1861*b*, pl. 16, fig. 12) is accurate. Only a single specimen, the holotype BRLSI M 41, remains in Moore's collection, but other specimens may be distributed in other collections.
Distribution. Westbury Formation, Beer Crocombe and east and west of Watchet, Somerset (Dawkins 1864).

Subclass EUTHYNEURA Spengel, 1881
Superorder OPISTHOBRANCHIA Milne-Edwards, 1848
Order CEPHALASPIDEA Fischer, 1880–87
Superfamily CYLINDROBULLINOIDEA Wenz, 1938–44
Family CYLINDROBULLINIDAE Wenz, 1938–44
Genus CYLINDROBULLINA von Ammon, 1878

Cylindrobullina elongata (Moore, 1861*b*)
Plate 10, figure 9

Description. BRLSI M 50, the holotype, is a small, cylindrical shell up to 6 mm high. Whorls flat sided, smooth. Shoulders of whorls pronounced, with ramps leading to an impressed suture, giving a gradate appearance to shell outline. Last whorl comprises two-thirds of total height. Shape of aperture unknown.

Remarks. More cylindrical in outline than other Penarth Group cylindrobullinids. According to Cossmann (1895, p. 161), *C. oviformis* and *C. ovalis* (see below) may be varieties of *C. elongata*. Examination of the specimens shows that *C. ovalis* is distinctly more globular and *C. oviformis* is biconical.

Distribution. Westbury Formation, Beer Crocombe, and east and west of Watchet, Somerset (Dawkins 1864).

Cylindrobullina oviformis (Moore, 1861*b*)
Plate 10, figure 10

Description. BRLSI M 51, the holotype, is a small, biconical shell 6 mm high. Early whorls which comprise the conical spire shallowly convex. Last whorl moderately convex, tapering markedly abapically. Whorls generally smooth with some faint growth lines visible. Suture slightly impressed, tops of whorls shouldered. Last whorl comprises more than three-quarters of total shell height. Aperture is a narrow, flattened oval (no longer visible on holotype due to mounting of specimen on card).

Remarks. The drawing in Moore (1861*b*, pl. 16, fig. 21) is distinctly more elongate than the examined specimen of this taxon, which has a distinctive biconically squat appearance.

Distribution. Westbury Formation, Beer Crocombe and Watchet, Somerset. Cotham Member, Pylle Hill, Bristol.

Cylindrobullina ovalis (Moore, 1861*b*)
Plate 10, figures 11–12

Description. BRLSI M 49, the holotype, is a small, globular shell, up to 9 mm high. Early whorls moderately convex, last whorl more so. Slight spiral band at the top of whorls, which are otherwise smooth. Suture impressed and shouldered. Last whorl comprises almost four-fifths of total shell height. Aperture slit-like, broadens abapically.

Remarks. The most globular of the Penarth Group cylindrobullinids.

Distribution. Westbury Formation, Beer Crocombe and Watchet, Somerset. Patchway, Bristol, horizon unknown.

Cylindrobullina fusiformis (Moore, 1861*b*)
Plate 10, figure 13

Description. BRLSI M 52, the holotype, is a tiny, fusiform shell 3 mm high. Adapertural sides of whorls moderately convex, adapical parts flat-sided with shouldered tops. Suture impressed, elevated spire gradate. Whorls smooth with no visible ornament. Last whorl comprises two-thirds of total height. Aperture narrow, pyriform.

Remarks. The whorls of the spire are flat-sided and the spire is much more gradate than that in the illustration in Moore (1861*b*, pl. 16, fig. 18).

Distribution. Westbury Formation, Beer Crocombe, Somerset.

Incertae sedis
Genus 'CHEMNITZIA'
'*Chemnitzia granum*' Dittmar, 1864
Plate 10, figure 14

Description. BRLSI M 42, the holotype, is a small, turreted shell up to 4 mm high. Whorls moderately convex, suture impressed. Whorls very smooth and polished, crossed by very faint axial ornament. No growth lines visible.

Remarks. Moore (1861*b*) gave the name *Chemnitzia nitida* (*non* Lycett) to specimens of this taxon from Beer Crocombe, Somerset. This was later amended by Hudleston and Wilson (1892). This taxon might possibly be a very abraded *Z. henrici*, but the true generic status is far from obvious.

Distribution. Westbury Formation, Beer Crocombe and Blue Anchor Point near Watchet, Somerset.

ADDITIONAL RECORDS

There are a few additional gastropod records from the Penarth Group, but these are identified to generic level only. Richardson (1911) recorded *Chemnitzia* spp. from the Westbury Beds (=Westbury Formation) near Watchet (Somerset) and *Pseudomelania*? and a depressed form of *Pleurotomaria* from the Langport Beds (=Langport Member) in the Charlton Mackrell railway cutting in Somerset. Finally, Ivimey-Cook (1975) recorded a species of *Actaeonina* (=*Cylindrobullina*?) in association with '*Natica*' *oppelii* (=*Naticopsina oppelii*) from the Lower Rhaetic (=Westbury Formation) of Antrim.

8. BIVALVES

by HUGH C. IVIMEY-COOK, PETER HODGES, ANDREW SWIFT *and* JONATHAN D. RADLEY

Marine bivalves are the most numerous and diverse of the macrofossils found in the Penarth Group, and occur throughout, except within some probably non-marine horizons in the Cotham Member. Non-marine bivalves are unknown. Little detailed taxonomic work on the bivalves has been published in this century although a number of revisions has been made to the genera in which previously described taxa had been placed. The following notes are intended to give an indication of the main characters of the taxa so that, with the illustrations, they facilitate their recognition; they are not presented as a definitive taxonomic revision.

A small number of bivalve genera, *Chlamys, Gervillella, Isocyprina* (*Eotrapezium*), *Modiolus, Protocardia* and *Rhaetavicula*, first appears in the uppermost horizons of the Blue Anchor Formation in south-west England, in beds recording the first marine influence of the Late Triassic transgression (Mayall 1981; Warrington and Ivimey-Cook 1995). The majority of the species described appeared gradually, to colonize the extensive marine habitats that became available during deposition of the Westbury Formation. Some taxa from the Westbury Formation persist into the Lilstock Formation, whilst others appear for the first time in that formation. Several of these species persist into the Jurassic.

Westbury Formation

Species of the bivalve genera *Chlamys, Isocyprina* (*Eotrapezium*), *Lyriomyophoria, Protocardia, Rhaetavicula* and *Tutcheria* are common and, in combination, are characteristic of the Westbury Formation. Species of between one and three of these genera tend to dominate separate bedding plane assemblages which were formed by wave or current-winnowing. These bivalve lumachelles are separated by less fossiliferous 'background' deposits in which bivalves are much scarcer. *Chlamys* commonly occurs in the more calcareous mudstones and in argillaceous limestones, sometimes accompanied by *Modiolus, Placunopsis* and fragments of *Liostrea*.

Many of the well-preserved specimens described and figured by Charles Moore (1861*b*) were attributed by him to the 'Flinty-bed' from near Beer-Crowcombe (now Beer Crocombe), south-east of Taunton, Somerset. This is a pale grey, shelly, limestone which he believed occurred just above the base of his 'Avicula contorta beds, Rhaetic Formation'; this bed is now placed low in the Westbury Formation. His specimens, however, came from dumps of material from excavations for a canal tunnel which is now disused and blocked. Today, only small fragments of this material can be found. Whilst many of these fragments have a consistent type of limestone lithology, Woodward (*in* Ussher 1906) was doubtful that all the specimens so labelled were from one horizon. He recorded a similar, but much less fossiliferous bed, 3·62–3·88 m above the base of the 'Black Shales' (Westbury Formation) at a nearby section in Hatch Beauchamp. Richardson (1911, pp. 6, 34, 52) indicated horizons that he considered equivalents of this bed but none has yielded faunas anything like as diverse, or with such well preserved specimens, as Moore's material from near Beer Crocombe.

Bivalves are commonly found disarticulated in bedding plane assemblages. The robust left valve of the characteristic Westbury Formation species *Rhaetavicula contorta* dominates very significantly over its more fragile right valve. Articulated bivalves (principally species of *Lyriomyophoria* and *Protocardia*) occur most commonly in the shales between these richer accumulations. At some localities around the Severn estuary, a series of depositional 'cycles' can be identified. At the base of each cycle is a silty or lutitic limestone, calcareous silt or fine sandy layer with shell detritus, which is succeeded by bivalve-rich mudstones. These mudstones become poorer in bivalves upwards towards the next coarse sediment layer (Ivimey-Cook 1974; Waters and Lawrence 1987, fig. 30). Some of the bivalves, such as *Chlamys* were probably nektic, whilst many others like *Modiolus* and *Rhaetavicula* were essentially epifaunal. *Isocyprina* (*Eotrapezium*), *Lyriomyophoria* and *Protocardia* and others may have been infaunal shallow burrowers.

Commonly the original calcite of the bivalve shells has been dissolved away to leave internal and external moulds. Disseminated pyrite was frequently deposited as a lining on the insides of valves as a result of the former presence of organic matter; this has commonly survived the removal of the calcareous shell material, giving rise to pyrite linings within external moulds.

Lilstock Formation
The lowest grey, calcareous mudstones of the Cotham Member locally include derived clasts of darker, fossiliferous, Westbury Formation shale, as around Penarth and Lavernock, South Glamorgan (Waters and Lawrence 1987). These clasts yield bivalves such as *Chlamys*, *Protocardia* and *Isocyprina* (*Eotrapezium*). These are accompanied by contemporaneous examples of the same genera, providing evidence for storm activity that reworked the underlying shelly muds. These calcareous mudstones between the top of the dark shales of the Westbury Formation and the pale calcareous grey-green mudstones of the Cotham Member yield the most diverse and prolific bivalve faunas in the Penarth Group.

Much of the overlying Cotham Member is poorly fossiliferous and bivalves are notably rare, although genera such as *Chlamys* occur sporadically in some pale grey, fine-grained limestones. Separate horizons in the green-grey, and locally red-tinged, marls yield either marine or non-marine macro- and microfossils; freshwater bivalves, however, are unknown. The stromatolitic Cotham or Landscape Marble near the top of the member may incorporate both fragmentary and complete bivalves, both within the algal layers or as accumulations between algal 'heads'. Specimens of *Modiolus* are locally abundant in the mud-flake breccias locally known as 'Crazy Cotham Marble', and are also found in splintery micrites associated with the channels between the algal bioherms.

The Langport Member is dominated by limestones; these can be massive and micritic and normally yield few macrofossils. However, aggregations of bivalves do occur locally in the limestones and occasionally in the intervening marls. Hallam (1960) and other authors have recorded a wide range of bivalve taxa, including arcids and myids, often preserved as moulds. The Langport Member is often condensed in marginal depositional settings, and bivalves frequently colonized these firmer substrates. The Pseudomonotis Bed in Gloucestershire (Richardson 1904) contains winnowed shell pavements of a species of *Oxytoma* previously placed in the genus *Pseudomonotis*, whilst other silty, cal-careous mudstones, limestones and fissile shale horizons may preserve *Atreta, Cardinia, Liostrea, Modiolus, Plagiostoma* and *Pteromya*.

In Northamptonshire and west Cambridgeshire, to the north of the Anglo-Brabant Massif, boreholes have proved a series of calcareous silts and limestones which are attributed to the Penarth Group. These are termed the Twyford Beds and have yielded a sparse fauna of Late Triassic marine bivalves (Horton *et al.* 1987; Warrington and Ivimey-Cook 1992).

Certain *Gastrochaenolites*-bored rockgrounds in the Carboniferous Limestone of the Mendip Hills may be of Rhaetian age, although at least some of these may also equate to Jurassic transgressive events.

The taxa reviewed below have been figured previously, some only as line drawings in the early literature, by authors such as Portlock (1843), Quenstedt (1858), Moore (1861*b*) and Etheridge (1872). Later, photo-graphs were used by Reynolds and Vaughan (1904), Richardson (1905), Richardson and Tutcher (1916), Arkell (1933) and Cox (1961, 1963) to illustrate further taxa. The majority of known taxa are illustrated and briefly discussed, with the author and date of original description of the species being given and listed in the references. The bivalve specimens illustrated by Moore (1861*b*, pls 15–16) in his fundamental systematic work have been specially photographed and most are re-figured herein. Comprehensive sections and lists outlining occurrences of Penarth Group bivalves at many localities have been provided by the extensive publica-tions of Linsdall Richardson (e.g. 1904, 1905, 1906, 1911). Later authors, such as Reynolds (1946) and Hallam (1960), and records in British

Geological Survey Memoirs (e.g. Whittaker and Green 1983; Waters and Lawrence 1987) have further extended knowledge of these faunas. A range chart illustrating the stratigraphical distribution of bivalves against lithologies in the Lavernock section (South Glamorgan) is given in Waters and Lawrence (1987, p. 73). Taxa currently thought to be first recorded from the basal Lias Group (the late Rhaetian Preplanorbis Beds of authors) are not included. However, it is to be hoped that publication of this guide will stimulate collecting so that more taxa are identified and the ranges of Penarth Group bivalves may be more accurately determined.

The classification of genera into higher taxonomic categories is based on Carter (1990) and Skelton and Benton (1993), but the lack of data on the internal characters of many taxa makes attributions to higher taxonomic units very tentative. None of the Penarth Group taxa, except *Rhaetavicula contorta*, appears to have been the subject of shell mineralogy studies such as those used by Carter (1990) in his reassessment of taxonomic relationships.

In the following descriptions small is used for less than 25 mm, medium for 25–50 mm and large for taxa exceeding 50 mm long.

Figured specimen repositories: BGS=British Geological Survey, Keyworth, Nottinghamshire; BRLSI=Bath Royal Literary and Scientific Institute, Bath; BMNH=The Natural History Museum, London; NMW=National Museum of Wales, Cardiff; SM=Sedgwick Museum, Cambridge; BRSUG=University of Bristol Geology Museum; AS=A. Swift collection, Department of Geology, University of Leicester.

SPECIES DESCRIPTIONS

Class BIVALVIA Linnaeus, 1758
Subclass PALAEOTAXODONTA Korobkhov, 1954
Order NUCULOIDA Dall, 1889*a*
Superfamily NUCULOIDEA Gray, 1824
Family NUCULIDAE Gray, 1824
Genus DACRYOMYA Agassiz, 1840
Dacryomya sp.
Plate 11, figures 1–2

Description. Small, umbones opisthogyrous, prominent. Shell outline dorsally concave with fairly smooth curve to rounded anterior end, posteriorly extended to form a short, almost pointed rostrum. More inflated than species of *Ryderia* (see below).
Distribution. Scarce, Westbury Formation.

Superfamily NUCULANOIDEA Adams and Adams, 1858
Family POLIDEVCIIDAE Kumpera, Prantl and Ruzicka, 1960

Genus RYDERIA Wilton, 1830
Ryderia titei (Moore, 1861*b*)
Plate 11, figures 3–4

Description. Small, up to 15 mm long, equivalve, highly inequilateral, umbones incurved over oblong area. Posterior region elongated with a narrow rostrum, anterior and ventral margins rounded. Valves moderately inflated, thick shelled. Ornament consists solely of commarginal striae.
Remarks. May be conspecific with *Ryderia doris* (d'Orbigny 1849). Plate 11, figure 5 shows an internal view of a specimen attributed to *Ryderia* sp. Specimens are often found with both valves closed in life position.
Distribution. Scarce, but occurs locally in small groups in the shales of the Westbury Formation.

Ryderia cf. *texturata* (Terquem and Piette, 1865)
Plate 11, figures 6–7

Description. Equivalve, inequilateral, with length (up to 30 mm) much greater than height. Outline club-shaped, posterior region elongated, tapering to form a prominent rostrum. Anterior margin rounded, valves moderately inflated. Umbones small, opisthogyrate and positioned anteriorly. Lunule long, slender. Commarginal ornament present on anterior part of shell, which changes close to and on the long narrow rostrum where it becomes sinuous, oblique and even vertical to the margin. Internal characters not known from specimens from the Penarth Group.
Distribution. Very rare, found in the upper Rhaetian (at a horizon possibly equivalent to the Langport Member) in Yorkshire (Gaunt *et al.* 1980). Becomes common in lowest Lias Group, ranges up to the *margaritatus* Biozone.

Subclass PTERIOMORPHIA Beurlen, 1944
Order ARCOIDA Stoliczka, 1871
Superfamily ARCOIDEA Lamarck, 1809

Remarks. Bivalves attributed to this superfamily have been recorded from the Penarth Group, and some are retained in museum collections. Virtually all are external moulds or casts in pale grey- or cream-coloured micritic limestone recorded as 'White Lias' (now Langport Member). Although it remains uncertain whether all of these are from the 'White Lias', rather than the higher 'Blue Lias Formation', there are certainly some from the Langport Member in Devon, for example from Pinhay Bay (Hallam 1960). Others are known from the thick Langport Member of Warwickshire.

The Triassic to Early Jurassic members of the Arcoidea were discussed by Arkell (1930). They were then divided amongst three families (Newell,

PLATE 11

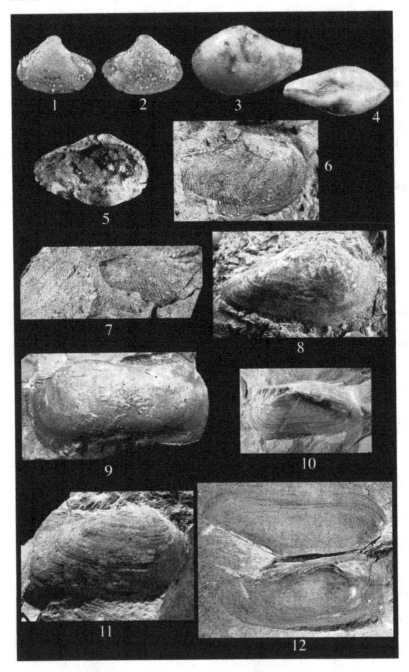

in Cox *et al.* 1969), partly on external shape and ornament but principally on the form of the ligamental area and hinge structure.

Arcidae have a hinge line with numerous small, more or less transverse, straight or chevron-shaped, unequal teeth diminishing in size towards the beak area. These teeth are radially disposed with respect to a point well below the beaks. The family includes *Arca* Linnaeus and *Barbatia* Gray.

Parallelodontidae have a long, straight, hinge line with strongly inequilateral teeth consisting of a few posterior laterals and numerous, generally short, oblique, horizontal or falcate, anterior and medial laterals. *Parallelodon* Meek and Worthen, *Grammatodon* Meek and Hayden and *Catella* Healey are examples of the genera found in Triassic–Early Jurassic strata (Arkell 1930).

Cucullaeidae have heavy shells with nearly median umbones and a complex hinge line. Their teeth include both transverse denticles and long symmetrically arranged anterior and posterior pseudolaterals. The oldest genus is *Cucullaea* Lamarck from the Lower Jurassic. There are no definite specimens of this family recorded from the Penarth Group.

Moore (1861*b*, p. 501, pl. 16, fig. 7) described and figured as '*Arca lycettii*' a new taxon from the 'Flinty-bed' (lower part of the Westbury

---

EXPLANATION OF PLATE 11

RV = right valve, LV = left valve

Figs 1–2. *Dacryomya* sp.; BGS AP 3526; Westbury Formation; Apley Barn Borehole, Witney, Oxfordshire, at *c.* 180 m. 1, internal mould, LV. 2, internal mould, RV. Both ×4.

Figs 3–4. *Ryderia titei* (Moore, 1861*b*); holotype BRLSI M 68; 'Flinty-bed', Westbury Formation; Beer Crocombe, Somerset. 3, LV. 4, dorsal view of both valves. Both ×3.

Fig. 5. *Ryderia* sp.; BGS AL 938; Westbury Formation; Wilkesley Borehole, Audlem, Cheshire at 163·98 m; internal, LV; ×4.

Figs 6–7. *Ryderia* cf. *texturata* (Terquem and Piette, 1865); BGS BLF 3784; Lilstock Formation; Scalla Moor Borehole, Yorkshire at 34·89 m. 6, external, RV. 7, shell fragments on core surface showing, to left, ornament on distal end of syphon and, to right, posterior dentition and proximal part of syphon. Both ×2.

Fig. 8. *Grammatodon* sp.; BRLSI M 69; 'Flinty-bed', Westbury Formation; Beer Crocombe, Somerset; external, RV; ×2.

Fig. 9. *Grammatodon* sp.; BGS GSM 90542; Lilstock Formation; between Lyme Regis and Axminster, Devon; internal, LV; ×2.

Figs 10–12. *Grammatodon* (*Cosmetodon*) *hettangiensis* (Terquem, 1855). 10, BMNH L 25347; Lilstock Formation; Garden Cliff, Westbury-on-Severn, Gloucestershire; RV. 11, NMW 46.311.G17; Lilstock Formation; quarry near Ely Racecourse, Cardiff, external; RV. 12, BMNH L 77168; Langport Member; Rockhill Quarry, Radstock, Somerset; both valves. All ×2.

Formation) at Beer Crocombe. He described the external features of the rare, single valves but did not see the interarea or hinge structure. Moore's (1861*b*) figured specimen, BRLSI M 69, is regarded here as *Grammatodon* sp. (Pl. 11, fig. 8), as is another specimen (Pl. 11, fig. 9). Other related taxa include *Grammatodon* (*Cosmetodon*) *hettangiensis* (Terquem) which is described below, and probably *Arca sinemuriensis* Martin, from the 'White Lias' of Stoneythorpe, Warwickshire. Richardson (1911) recorded '*Grammatodon lycetti* Moore' from the Langport Member of Langport, Somerset.

Family PARALLELODONTIDAE Dall, 1898
Subfamily GRAMMATODONTINAE Branson, 1942
Genus GRAMMATODON Meek and Hayden, 1861
Subgenus COSMETODON Branson, 1942
Grammatodon (*Cosmetodon*) *hettangiensis* (Terquem, 1855)
Plate 11, figures 10–12

Description. Small to medium sized, equivalve, inequilateral, ovate-elongate to trapezoidal in outline, valves well inflated, anterior margin rounded, meeting dorsal margin at a right angle. Posterior margin straight, meeting dorsal margin at an obtuse angle. Ventral margin relatively straight with slight concavity at the central part of the margin. Valve margins closed. Umbones positioned somewhat anteriorly, salient above straight hinge line, slightly opisthogyrate. Weak anterior and stronger posterior ridge from umbo to margins. Ornament of commarginal growth lines and numerous fine, radial ribs.
Distribution. Lilstock Formation, near Ely Racecourse, Cardiff; Langport Member, Garden Cliff, Gloucestershire; Wimpston, Warwickshire; Rockhill Quarry, Radstock, Somerset; Pinhay Bay, Devon.

Order PTERIOIDA Newell, 1965
Suborder PTERIINA Newell, 1965
Superfamily PTERIOIDEA Gray, 1847
Family PTERIIDAE Gray, 1847
Genus RHAETAVICULA Cox, 1961
Rhaetavicula contorta (Portlock, 1843)
Plate 12, figures 1–3

Description. Small, commonly up to 12 mm but occasionally up to about 25 mm long. Narrow, lunate, strongly prosocline, highly inequivalve, alate. Valves discordant with left valve strongly convex, posterior margin upturned and ornamented with unevenly spaced, relatively numerous, slightly irregular ribs between which interdigitating radial threads appear towards posterior. Shell can be twisted, and may be arched up over hingeline. Umbo salient, anteriorly placed. Left anterior wing very small; posterior wing usually long, pointed and ornamented with growth lines. Right valve nearly flat, thin, smaller than left, with a deep rounded sinus, fine growth lines and only very fine ribs. Anterior auricle small, high in proportion to length and not separated by a notch. Posterior auricle variable, but may be as long as valve itself.

Remarks. The characteristic bivalve of the European Rhaetian, referred in the past to the genera *Avicula, Cassianella* and *Pteria*, in Britain is often abundant in the Westbury Formation (the Contorta Shales of nineteenth century authors) but, with the recent restriction of the Westbury Formation to the dominantly non-calcareous grey mudstones facies (Warrington *et al.* 1980), now also ranges up into the lowest part of the calcareous silts of the Cotham Member. Normally, only the thicker left valve is preserved intact; however, even fragments may be easily recognized by the solid ribs, which are almost round in cross section and rarely straight. The right valve fits within the left, is much thinner and is much less frequently preserved but can be found in current-sorted assemblages. The holotype (Plate 12, fig. 1), from near Magilligan, Northern Ireland, is rather poorly preserved in a coarse siltstone (Portlock 1843). In the Westbury Formation *R. contorta* is locally abundant both as bedding plane assemblages in association with a few other taxa and also as comminuted fragments in more silty and calcareous horizons. When Cox (1961) proposed the genus *Rhaetavicula* he also reviewed the classification of this species and pointed out its taxonomic similarity to Oxytomidae. However, Bøggild (1930) considered that the originally calcitic outer shell layer was essentially structureless, whereas the inner shell layers (presumed to be aragonitic) were not preserved. Carter (1990, p. 206) preferred to place the genus in Pteriidae.

Vaughan (1904, pp. 203–204) concluded that small specimens of this species 'exactly resemble *Avicula solitaria* Moore, 1861*b* (pl. 15, fig. 11), in the absence of intercalation'. Following Cox (1961) the species *Avicula solitaria* Moore, 1861*b* is rejected as an independent taxon, being considered to be an immature form of *R. contorta*.

Distribution. Common in the Westbury Formation. Examples from the base of the Cotham Member include both reworked and contemporaneous shells but it did not persist for long in this facies and is unknown from the Langport Member.

Family BAKEVELLIIDAE King, 1850
Genus GERVILLELLA Waagen, 1907
Gervillella praecursor (Quenstedt, 1856)
Plate 12, figures 4–5

Description. Small, up to about 20 mm but usually around 10 mm long, thin shelled. Oblique, longitudinally elongated, slightly inflated but usually

found crushed flat. Anterior margin convex without excavation. Surface ornamented with relatively prominent growth lines and undulations.
Remarks. Internal moulds show hinge lines of bakevelliid affinity. Moore (1861*b*, p. 500) noted what he believed to be original colour banding on his specimens, with six or seven broad, longitudinal bands of a pinkish colour.
Distribution. Found throughout the Penarth Group, particularly common towards the top of the Westbury Formation and in the base of the Cotham Member, sporadic above this in the Lilstock Formation.

'*Gervillia*' *ornata* Moore, 1861*b*
Plate 12, figure 6

Description. Similar in shape to *G. praecursor*, but significantly smaller. Surface ornamented with six or seven prominent longitudinal ribs, extending from umbo to ventral margin. Ribs crossed by raised growth lines. Details of hinge line unknown.
Remarks. Only three specimens are known, all from Moore's 'Flinty-bed'

EXPLANATION OF PLATE 12

RV=right valve, LV=left valve, cop.=copy of

Figs 1–3. *Rhaetavicula contorta* (Portlock, 1843); Westbury Formation. 1, cast of holotype BGS GSM 7821; Magilligan/Aghanloo, Co. Londonderry; LV. 2, BGS GSM 100623; Stowell Park Borehole, Northleach, Gloucestershire at 633·05 m.; RV. 3, BGS BKB 6669; Great Orton No. 1 Borehole, Cumberland at 90·85 m.; LV. All ×2.

Figs 4–5. *Gervillella praecursor* (Quenstedt, 1856); 'Flinty-bed', Westbury Formation; Beer Crocombe, Somerset. 4, BRLSI M 76; LV. 5, BRLSI M 74; RV. Both ×3.

Fig. 6. '*Gervillia*' *ornata* Moore, 1861*b*; BRLSI M 80; 'Flinty-bed', Westbury Formation; Beer Crocombe, Somerset; LV; ×3.

Figs 7–8. *Plagiostoma giganteum* J. Sowerby, 1814: 7, SM J 58595; Lilstock Formation; Penarth, Glamorgan; LV. 8, cop. Richardson 1905, pl. 33, fig. 3 as *Lima valoniensis*, Langport Member; Coldknap, Barry, Glamorgan; LV. Both ×1·5.

Figs 9–10. *Plagiostoma punctatum* J. Sowerby, 1815. 9, BRLSI M 117; 'Flinty-bed', Westbury Formation; Beer Crocombe, Somerset; external, RV. 10, BGS ×4735; Penarth Group; Penarth; external, RV. Both ×1·5.

Fig. 11. *Plicatula cloacina* Vaughan, 1904; composite drawing from fragments; cop. Vaughan 1904, p. 202, text-fig. 3; 'Bone Bed', Westbury Formation; Gloucestershire; 'upper valve'; ×2.

Fig. 12. *Plicatula hettangiensis* Terquem, 1855; cop. Richardson 1905, pl. 33, fig. 2; Langport Member; Coldknap, Barry, Glamorgan; RV; ×1·5.

PLATE 12

collection (see Moore 1861*b*, pl. 15, fig. 8). The taxon may be related to *Gervillia*, but this is not established.

Distribution. Only known from the 'Flinty-bed' of the basal Westbury Formation at Beer Crocombe, Somerset.

Order LIMOIDA Rafinesque, 1815
Superfamily LIMOIDEA Rafinesque, 1815
Family LIMIDAE Rafinesque, 1815
Genus PLAGIOSTOMA J. Sowerby, 1814

Description. 'Medium size to large, obliquely ovate, opisthocline, length commonly slightly exceeds height, inflation moderate to strong; beaks more or less anterior to the middle of a moderately long cardinal area; ligament pit broad; auricles obtuse; anterior umbonal ridge well defined with excavated lunule in front of it; marginal gapes small or absent; edentulous or with one or two broad, longitudinally directed teeth in each dorsal angle; surface smooth, radially striated, or with weak ribs commonly with punctate intervals' (Cox, *in* Cox *et al*. 1969, p. N391). *Plagiostoma giganteum* J. Sowerby is the genotype.

Remarks. Two species of this genus are recorded from the Penarth Group. Richardson (1905, p. 423) revived the taxon *Lima valoniensis* Defrance, first described and figured by de Caumont (1825, p. 507, Atlas pl. 22, fig. 7) and later redefined by Dumortier (1864). Richardson (1905) used this species for examples obtained principally from the Langport Member in Glamorgan, but also added an observation by Vaughan (Richardson 1905, pp. 423–424), suggesting that this 'typical clavellate form' also occurred in the 'main Pecten Bed' at Lilliput, Chipping Sodbury, which is near the top of the Westbury Formation. However, Vaughan (1904, p. 202) recorded only fragments of *Lima* sp. in this bed.

This taxon appears distinct from that described and figured by Moore (1861*b*, p. 500, pl. 15, fig. 9) which he attributed to *Lima praecursor* Quenstedt, 1856 (p. 29, pl. 1, figs 22–24), despite Quenstedt's own reservations about the separateness of his taxon from *Plagiostoma giganteum* J. Sowerby, 1814 (see below).

Whilst two species are thought to occur in the Penarth Group, both are here attributed to James Sowerby's species and the taxa *Lima valoniensis* (Defrance, *in* de Caumont 1825) and *Lima praecursor* Quenstedt, 1856, as used by Moore (1861*b*), are considered synonyms of *Plagiostoma giganteum* J. Sowerby, 1814 and *Plagiostoma punctatum* J. Sowerby, 1815 respectively.

Plagiostoma giganteum J. Sowerby, 1814
Plate 12, figures 7–8

Description. Equivalve, inequilateral, oblique; outline sub-ovate to slightly deltoid. Height slightly less than length. Moderately inflated. Ventral and posterior margins rounded, antero-dorsal margin straight. Beaks positioned towards posterior, slightly opisthogyrate, salient above hinge line. Valves bi-auriculate, with posterior auricle more pronounced than anterior, the latter usually depressed within a well-defined lunule. Shell surface smooth, porcellaneous. Near to posterior and anterior margins radial ribs slightly sinuous, irregular in size, very flattened. Extent of rib development often dependent on degree of wear of shell surface. Fine commarginal growth halts present, often pronounced. Worn shells show underlying layer to consist of fine, fibrous, divaricating striae. Internally hinge length moderate, with sub-triangular cardinal area and well-defined triangular ligament pit, crura absent. Hinge plane acutely angled to the plane of commisure in both valves. Musculature not seen.

Remarks. Sowerby's two figures (1814, pl. 77) show material typical of many specimens from the lower beds of the Lias Group. They resemble the specimens found in the Langport Member by Richardson and other collectors, who used the name *Lima valoniensis* Defrance. Langport Member specimens are usually considerably less than 40 mm long, much smaller than Lower Jurassic examples. Richardson (1905, p. 423, pl. 33, fig. 3) illustrated a specimen about 26 mm long (Pl. 12, fig. 8 herein). This latter is a weakly inflated form with a rather narrow sub-terminal umbo and weak ribbing below the umbo which strengthens peripherally into slightly sinuous, flat-topped strands which are characteristic of *Plagiostoma giganteum*.

Distribution. Although Vaughan (*in* Richardson 1905, p. 424) is quoted as recording the taxon from the 'main Pecten Bed' of the Westbury Formation, it is principally known from the Langport Member.

Plagiostoma punctatum J. Sowerby, 1815
Plate 12, figures 9–10

Description. Equivalve, inequilateral, oblique, sub-ovate to slightly ovoid in outline. Height slightly less than length and moderately inflated, attains 96×98 mm in size but the type specimens are half this size. Posterior and ventral margins rounded, anterio-dorsal margin straight. Beaks posterior, slightly opisthogyrate and salient. Valves bi-auriculate, posterior auricle slightly larger than anterior. Shell surface, including posterior auricle, ornamented with approximately 100 radial ribs, rounded in cross section but often flattened by wear. Ribs near postero-dorsal margin more angular. Fine, regularly spaced commarginal striae (30 counted in 5 mm at 30 mm from umbo) occur on shell surface, with greatest number near umbo. When worn this ornament is lost except in the narrow interspaces where it appears as regular rows of punctae.

Internally, hinge length moderate, cardinal area with well developed ligament pit, each dorsal angle with a single cardinal crus parallel to dorsal margin.

Remarks. The specimen illustrated on Plate 12, figure 9 is considered to belong to this taxon despite being very worn; however, it still possesses a small area of ribbing. It was figured by Moore (1861*b*, pl. 15, fig. 9) as *Lima praecursor* Quenstedt. J. Sowerby's specimens (1815, pl. 113, figs 1–2) in The Natural History Museum (4328a–b), as published, are mirror images of the types. The species differs from *P. giganteum* (Sowerby, 1814, p. 176, pl. 77) in having more pronounced radial ribs and regular commarginal striae which wear down to reveal regular rows of punctae in the interspaces (Pl. 12, fig. 10). *P. punctatum* does not attain the size of *P. giganteum* in the lower part of the Lias Group, where it is also common.

Specimens of *P. punctatum* have been collected from the Langport Member of the Bristol area (specimens Cf. 2793–2797 are held in Bristol City Museum and Art Gallery), but these are not well preserved.

Distribution. Moore's (1861*b*) specimen from the 'Flinty-bed' at Beer Crocombe is the sole example of this species recovered from the Westbury Formation. It is recorded from the Langport Member and persists into the Lower Jurassic, where it attains greater size.

Order OSTREOIDA Férussac, 1822
Superfamily PLICATULOIDEA Watson, 1930
Family PLICATULIDAE Watson, 1930
Genus PLICATULA Lamarck, 1801
Plicatula cloacina Vaughan, 1904
Plate 12, figure 11

Description. 'Upper valve strongly convex, narrowing towards the beak. Length up to *c.* 20 mm. Shell thin, composed of slightly overlapping concentric bands which have free ragged edges. Fine, sharp radial ribs cross these bands at irregular intervals and end on the free edges in projecting points. The outermost bands are ornamented with fine, close, parallel, concentric, rounded striae. To the left of the valve, the radial ribs and spines are much more numerous and strongly marked, several of the ribs are continuous and the spines closely packed, short and tubular. Lower valve very imperfectly known' (Vaughan 1904, p. 202).

Remarks. Vaughan (1904) erected this species on the basis of numerous shell fragments. However, neither the location of this original material, nor further specimens, are known to the authors. Vaughan's photograph (pl. 18, fig. 5) is of an indeterminate fragment; thus his description and text-figure are the only guides to the identity of the taxon, and these are somewhat contradictory. *P. cloacina* apparently differs significantly from *P.*

hettangiensis in both shape and ornament and also lacks the irregular valve outline and shell rugosity typical of other species of *Plicatula*. However, its radial ribs do appear to be developed into projecting points leading to short, tubular spines, which is a common feature in other *Plicatula* species. Whilst Vaughan (1904) included these features in his description, they are not shown in his text-figure 3. Further material is needed to interpret this taxon.

Distribution. Westbury Formation. Vaughan (1904) recorded *P. cloacina* principally from the 'bone bed' at the base of the Westbury Formation at a locality 200 m west of Chipping Sodbury Station.

Plicatula hettangiensis Terquem, 1855
Plate 12, figure 12

Description. Irregularly triangular, inequivalve, sub-convex. May be contorted to give both valves a wavy appearance in profile. Length up to 65 mm. Attachment area small. Small spines occur at intersection of roughly commarginal growth lines and a number of rugosities, which radiate from umbo to margins. Spines increase in size towards margin where they become tubular and may be more than 1 mm long. Some interdigitating rugosities between primaries. Superior valve often flattened, but may possess upcurved margin.

Remarks. Richardson (1905) expressed surprise that this species had not previously been described from the 'White Lias', as he found it to be common at Coldknap, Barry and Lavernock. He referred to variations in form and figured a specimen from Coldknap (Richardson 1905, pl. 33, fig. 2; reproduced here as Pl. 12, fig. 12) which is 27 mm long. Later, Richardson (1911) recorded *Plicatula hettangiensis* from the base of the Langport Member in a railway cutting near Dunball, Somerset.

Distribution. Langport Member, ranges into the Lias Group.

Plicatula spp.
Plate 13, figures 1–2

Remarks. Specimens occur quite commonly which are not distinctive nor well enough preserved to be identified specifically within *Plicatula*. A specimen from the Cotham Member of the Wilton Farm Borehole, near Northallerton, Yorkshire is shown as Plate 13, figure 1. Others are recorded from the upper, silty, part of the Langport Member at Lavernock, formerly attributed to the 'Watchet Beds' (Richardson 1911; Waters and Lawrence 1987). The genus is also recorded from the Langport Member at Burnet Hill, Keynsham, Bristol (Pl. 13, fig. 2) and from near Dunball, Somerset.

Superfamily DIMYOIDEA Fischer, 1886
Family DIMYIDAE Fischer, 1886
Genus ATRETA Étallon, 1862
Atreta intusstriata (Emmrich, 1853)
Plate 13, figures 3–4

Description. Small, height to about 12 mm, width to 9 mm, sub-orbicular to sub-ovate, but shape variable, may be irregular. Inequivalve and inequilateral. Cemented right valve attached by most of its surface area. Left valve follows contours of right valve, gently bowl-shaped, ornamented with fine, slightly sinuous radial imbrications. Left valve internally shows dimyarian musculature, with anterior adductor and bilobate posterior adductor. Right valve shallowly concave internally, with raised peripheral rim. Inner shell layer often missing, exposing inner surface of outer shell layer ornamented with anastomosing fine riblets with up to four orders of bifurcation.

Remarks. This taxon has been referred to a number of genera, including *Ostrea* (Emmrich 1853), *Spondylus* (Terquem 1855), *Plicatula* (Terquem and Piette 1865), *Dimyopsis* (Goetel 1916) and *Dimyodon* (Arkell 1933). Such uncertainty has resulted partly from difficulties in interpreting the internal structure. Hodges (1991) reviewed this species and, following Cox *et al.* (1969), considered *Dimyopsis* to be a junior synonym of *Atreta*, despite the considerable stratigraphical separation (Sinemurian to Callovian) from the type species of *Atreta* (*Ostrea blandina* d'Orbigny, 1849).

Distribution. Rare in the Westbury Formation; more common in the Lilstock Formation, often attached to hard substrates. Widespread in the Rhaetian in Europe and ranges into the Lias Group.

Superfamily OSTREOIDEA Rafinesque, 1815
Family OSTREIDAE Rafinesque, 1815
Subfamily OSTREINAE Rafinesque, 1815
Genus LIOSTREA Douvillé, 1904

Remarks. The genus *Liostrea* was established by Schäfle (1929) who placed many taxa of previously described ostreids into synonymy with *Liostrea irregularis* Münster (*in* Goldfuss 1833), which he considered had priority. He included amongst these synonyms *Ostrea hisingeri* Nilsson, *in* Lundgren, 1878, but did not point out that Nilsson's species was erected in 1832 and therefore had priority over Münster's species. The priority of *Liostrea hisingeri*, and also the synonymy with it of *Ostrea liassica* Strickland, 1845 is now widely accepted.

The extent of phenotypic variation which can be encompassed within a species has long been a problem in the ostreid complex. The presence of two species in the Penarth Group has been accepted for almost a century and, on concensus, this practice is maintained here, although one of us (PH) considers that *L. bristovi* is better regarded as a junior synonym of *L. hisingeri*.

Liostrea bristovi (Richardson, 1905)
Plate 13, figures 5–6

Description. Large, up to about 60 mm long and 55 mm wide. Shape variable but generally conforms to an ovate triangle. Valves flat to slightly convex, irregularly rounded, anterior and posterior margins converge regularly to the beak area. Shell fairly thin, attachment area probably quite small but rarely seen. Surface ornamented with some broad, irregular, concentric undulations and growth lines.

Remarks. Richardson (1905, p. 422, pl. 33, fig. 4; reproduced here as Pl. 13, fig. 6) followed Etheridge (1872) in considering this a new species, but he was the first to publish a description. *L. bristovi* is only known from the 'Sully Beds' of Glamorgan (Richardson 1905), which are beds, chiefly of limestone, below the base of the Lower Rhaetic (*sensu* Richardson 1905) in south-east Glamorgan. They are best seen in St Mary's Well Bay, also locally inland, where *L. bristovi* occurs in a limestone lumachelle. The base of the Westbury Formation is now placed at a non-sequence just below the base of these beds (Ivimey-Cook 1974).

Adult *L. bristovi* are thin-shelled, yet substantially larger than specimens of *L. hisingeri*. The latter are thicker-shelled, show marked concentric rugosity and possess a generally smaller attachment area. *L. bristovi* is also dissimilar to those *Liostrea* with large attachment areas, which some authors distinguish as *Liostrea irregularis* (Münster).

Distribution. Only known from the basal beds of the Westbury Formation in east Glamorgan.

Liostrea hisingeri (Nilsson, 1832)
Plate 13, figure 7

Description. Generally small, up to 25 mm long, generally ovate to pear-shaped, but variable when growth is modified by neighbouring shells. Irregularly rounded, shell flat to slightly convex. Growth lines generally commarginal. Rugose corrugations also generally follow the outline. Attachment area mostly small, superior valve more planar than underlying valve, which may take the shape of the attachment surface, sometimes producing a xenomorphic shape from a dead shell.

Distribution. *Liostrea hisingeri* is common in the lowest beds of the Lias Group. Small oysters, probably referable to *L. hisingeri* rather than to *L. bristovi*, occur sporadically in the Westbury Formation and more

PLATE 13

abundantly in the Langport Member, especially from central Somerset (Richardson 1911).

Subfamily LOPHINAE Vyalov, 1936
Genus LOPHA Röding, 1798
Subgenus ACTINOSTREON Bayle, 1878?
Lopha (*Actinostreon?*) *haidingeriana* Emmrich, 1853
Plate 13, figure 8

Description. Shell medium to large, thin, with little secondary thickening. Subequivalve, variable in size. Attachment area posterior to left umbo. Valve high, slightly extended, shape variable. Umbo not clear, probably recurved and ribbed to extremity. Sharp central ridge separates valve into a lateral and anterior portion. Four centro-lateral ribs arise by insertion in interspaces, but split into two on reaching extension of umbonal ridge, then fan out to margin. All posterior ribs arise from central rib and do not split. Anterior and lateral posterior ribs sharp, others rounded (based on Willey 1966).
Remarks. The figured specimen is attached to a large *Modiolus* shell.
Distribution. One record only, from grey, silty limestones of late Rhaetian age (?Lilstock Formation) near the southern end of Cowbridge railway cutting, Glamorgan (Willey 1966).

EXPLANATION OF PLATE 13

RV=right valve, LV=left valve, cop.=copy of

Figs 1–2. *Plicatula* sp. 1, BGS HIC 8281; Cotham Member; Wilton Manor Borehole at 14·17 m, Brompton, North Yorkshire; RV. 2, BGS Zn 4116; Langport Member; top of Burnet Hill, 3 km south of Keynsham, Bristol; LV. Both ×1·5.

Figs 3–4. *Atreta intusstriata* (Emmrich, 1853). 3, BGS GSM 90557; Penarth Group; Stockton, Warwickshire; internal of RV. 4, cop. Richardson 1905, pl. 33, fig. 1, as *Plicatula intus-striata* Emmrich; Langport Member; Coldknap, Barry, Glamorgan; LV. Both ×2.

Figs 5–6. *Liostrea bristovi* (Richardson, 1905); Westbury Formation; St Mary's Well Bay, Sully, Glamorgan. 5, NMW 22·345·G35; LV. 6, cop. Richardson 1905, pl. 33, fig. 4 as *Ostrea bristovi* 'Etheridge MS' ex. 'Sully Beds'; ?RV. Both ×1.

Fig. 7. *Liostrea hisingeri* (Nilsson, 1832); BGS JP 3054; Langport Member; railway cutting west of Priory, Charlton, Somerset; LV; ×1.

Fig. 8. *Lopha* (*Actinostreon?*) *haidingeriana* Emmrich, 1853; NMW 67·121·G4; near base of Lilstock Formation; south end of Cowbridge railway cutting, Glamorgan; RV; ×1.

Order PECTINOIDA Rafinesque, 1815
Superfamily MONOTOIDEA Fischer, 1887
Family OXYTOMIDAE Ichikawa, 1958
Genus OXYTOMA Meek, 1864
Oxytoma fallax (Pflücker, 1868)
Plate 14, figures 1–2

Description. Small, up to 7 mm long, rounded to sub-orbicular, inflated, inequivalve. Umbo slightly protruding. Left valve more strongly convex than right. Hinge line straight, posterior auricle not separated by sulcus from flank. Ornamented with about 24–26 conspicuous radial riblets.

Remarks. This species is distinguished from *Meleagrinella*, with which it was previously classified, by the presence of pronounced matching anterior auricles on both valves. It has much in common with *Oxytoma* (*Hypoxytoma*) Ichikawa, 1958, although this subgenus has not previously been recorded below the Lower Cretaceous. Another subgenus, *Palmoxytoma* Cox, 1961, occurs quite widely in the lowest beds of the succeeding Lias Group. '*Meleagrinella*' *decussata* Münster, *in* Goldfuss, 1831, which occurs in the lower beds of the Lias Group, is most readily distinguished from *O. fallax* by having a greater number (36–44) of radial ribs and radial ornamentation on the auricles. These two taxa are often confused in records.

The taxon has been previously attributed to *Avicula*, *Monotis* and, later, to *Pseudomonotis*, hence the term Pseudomonotis Bed for a horizon where it is notably abundant.

Distribution. Occasionally recorded from the Westbury Formation. Richardson (1911, p. 7) cited it as common in the Cotham Marble near the top of the Cotham Member. Abundant in the cream-coloured, fissile Pseudomonotis Bed in Gloucestershire (Richardson 1904). Although this bed may have been deposited at the same time as some part of the Langport Member elsewhere it could also be considered, on lithological grounds, as the lowest part of the Lias Group in the Severn Basin.

Family TERQUEMIIDAE Cox, 1964
Genus TERQUEMIA Tate, 1867*b*
Terquemia difformis (Schlotheim, 1820)
Plate 14, figure 3

Description. Medium–large in size, shape variable. Valves may be cemented together in clusters. Inequivalve, inequilateral, sub-oval to sub-orbicular, often irregular. Shell oblique, both valves slightly higher than long; right valve strongly convex, left valve flattened to slightly convex. Medium to large attachment area on right valve often takes a xenomorphic impression of other shells, area occupies 70 per cent. of shell surface of some specimens. The remaining, unattached, part of right valve, and left valve, have ornament of numerous strong, rounded, radial, sometimes sinuous ribs with maximum amplitude distant from hinge line where they become acutely angled. Ribs may bifurcate with commissural plications at shell margins. Commarginal striae may be present, as may pronounced growth lines.

Remarks. The figured specimen had previously been assigned to *Lima terquemi* and was obtained from an unspecified Rhaetian horizon.

Distribution. Rare, but has been recorded from the Langport Member in Somerset. The taxon persists well into the Lias Group.

Superfamily ANOMIOIDEA Rafinesque, 1815
Family ANOMIIDAE Rafinesque, 1815
Genus PLACUNOPSIS Morris and Lycett, 1853
Placunopsis alpina (Winkler, 1859)
Plate 14, figures 4–6

Description. Medium-sized, up to about 40 mm long, thin-shelled, convex and variably rounded. Hinge line short, nearly straight. Beak small, slightly projecting, sometimes with a slight ear-like projection on posterior side. Smaller valve flattened to concave, close fitting. Surface ornament variable from fairly smooth with fine commarginal growth lines where external shell layers are preserved, to strong concentric wrinkles. Slight erosion of this surface exhibits numerous close-set, slightly sinuous, longitudinal striae (Pl. 14, fig. 5).

Remarks. Often found associated with *Chlamys valoniensis* in the more calcareous beds of the Westbury Formation.

Distribution. Principally found in the Westbury Formation, also from the basal beds of the Cotham Member.

Superfamily PECTINOIDEA Rafinesque, 1815
Family PECTINIDAE Rafinesque, 1815
Genus CAMPTONECTES Agassiz, *in* Meek, 1864
Camptonectes sp.
Plate 14, figure 7

Description. Equivalve, left valve more convex than right. Large byssal notch. The concentric sculpture of fine growth lines may be crossed by fine, oblique, curved, crenulate or commonly punctate striae. Some taxa show concentric raised laminae.

Remarks. This is a very rare form. Johnson (1984) recorded the lowest occurrence of this genus from the Preplanorbis Beds (lowermost Lias Group), but there are occasional records of probable Penarth Group examples such as the specimen figured here. Rare references to *Entolium*

PLATE 14

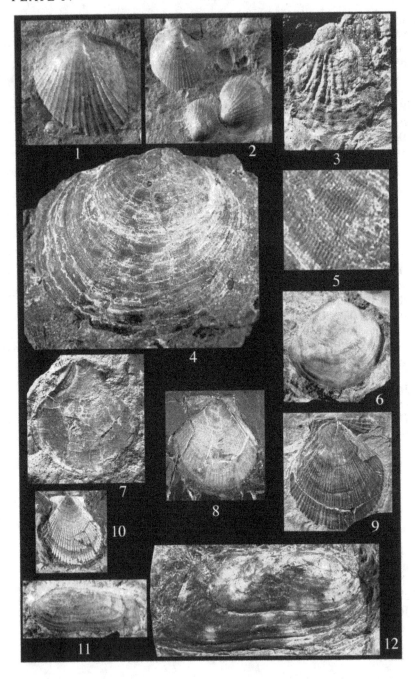

Meek, 1865, a smooth pectinid with no byssal notch and subequal auricles, and *Syncyclonema* Meek, 1864, are probably misidentifications of either taxon or horizon.

Genus CHLAMYS Röding, 1798
Chlamys valoniensis (Defrance, *in* Caumont, 1825)
Plate 14, figures 8–10

Description. Medium- to large-sized, length generally up to 40 mm, exceptionally up to 63 mm (Johnson 1984). Slightly inequilateral with a somewhat larger posterior part, sub-orbicular, bialate. Equivalve, relatively compressed, but inflation of left valve may be double that of right. Umbonal angle approximately 90°. Auricles well-marked, unequal, posterior auricle with about five fine ribs. Valves covered with fine, quite closely spaced ribs or plicae; early in ontogeny these may number 20–30. Right valve has slightly fewer ribs than left (Johnson 1984). Ribs raised at intersections with fine commarginal growth lines. Irregularly intercalated secondary ribs occur in large specimens.

Remarks. This taxon was attributed to 'Defrance' by de Caumont (1825, p. 507) and has been attributed in this way for over 150 years. The type

---

EXPLANATION OF PLATE 14

RV=right valve, LV=left valve

Figs 1–2. *Oxytoma fallax* (Pflücker, 1868); AS 82/2; Pseudomonotis Bed; Westbury on Severn. 1, LV; ×4. 2, group of three left valves; ×2.

Fig. 3. *Terquemia difformis* (Schlotheim, 1820); NMW 22·345·G443; Rhaetian; locality unknown; LV; ×1·5.

Figs 4–6. *Placunopsis alpina* (Winkler, 1859); Westbury Formation. 4, BGS HIC 6386; St Mary's Well Bay, Glamorgan, RV; ×1·5. 5, same specimen; ×4, to show finely sinuous ornament. 6, BRLSI M 122a; 'Flinty-bed'; Beer Crocombe, Somerset; LV; ×1·5.

Fig. 7. *Camptonectes* sp.; NMW 89·64G·3; Rhaetian, Penarth Group?; Charlton Mackrell, Somerset; internal, RV; ×1·5.

Figs 8–10. *Chlamys valoniensis* (Defrance, 1825); Westbury Formation. 8, BGS Bj 6098; Stowell Park Borehole, Northleach, Gloucestershire at 631·97 m; internal, RV. 9, BGS Bc 4938; Saleway Borehole, Droitwich at 24·08 m; external, RV. 10, BGS Zf 1977; Uphill, Weston-super-mare, Somerset; LV. All ×1·5.

Figs 11–12. *'Permophorus' elongatus* (Moore, 1861*b*); Westbury Formation. 11, syntype, BRLSI M 146; 'Flinty-bed'; Beer Crocombe, Somerset; external, LV. 12, BGS BDE 6794; Burton Row Borehole, Somerset, at 415·07 m; external, RV. Both ×1·5.

species was reputedly from the 'Calcaire de Valognes', which is of Hettangian age and also yielded *'Ammonites eruca'*. The species was first well figured and described by Michelin (*in* Leymerie 1838). The type material was destroyed in Caen during the 1939–45 war.

Another form with an umbonal angle of 80–85° and fine sharper radial ribs also occurs and may be distinct but is currently unnamed. This form may be that described and figured by Vaughan (1904, pl. 18, fig. 2), which is both narrower and has a smaller umbonal angle than the usual form which he figured as plate 18, figure 2A.

Distribution. Topmost beds of the Blue Anchor Formation (Warrington and Whittaker 1984) and throughout the Penarth Group. It is locally abundant in the silty limestones and 'Pecten' beds of the Westbury Formation, but scarce in the Lilstock Formation. Ranges into beds of Hettangian age, where it is also most common in more silty or sandy beds (Johnson 1984).

Subclass ISOFILIBRANCHIA Iredale, 1939
Order MODIOMORPHOIDA Newell, 1969
Superfamily KALENTEROIDEA Marwick, 1953
Family PERMOPHORIDAE van de Poel, 1959
Subfamily PERMOPHORINAE van de Poel, 1959
Genus 'PERMOPHORUS' Chavan, 1954
'Permophorus' elongatus (Moore, 1861*b*)
Plate 14, figures 11–12

Description. Ovate–oblong, medium-sized, up to about 50 mm long, inequivalve. Rounded anterior margin, flattened posterior margin. Beaks anterior, weak. Two poorly to moderately well-developed lines, extending from the umbones to the postero-ventral margin, may divide the shell into three elongate triangular areas. Shell surface covered by fine growth lines. Hinge undescribed, Vaughan (1904, p. 205) described the interior of specimens as having 'a prominent anterior muscular impression, in front of the beak, circumscribed by a deep furrow, a pallial line of continuous curvature, ending in a less prominent posterior muscular impression, a very blunt beak region, a sharp indentation in front of the beak, continuous with the deep groove which forms the hinder boundary of the anterior muscular impression'. Vaughan concluded that these structures were appropriate to this 'genus'.

Remarks. The generic identity of this taxon is unclear. The genus *Pleurophorus*, to which it was earlier assigned, is preoccupied. It is rarely found other than as crushed casts and moulds.

Distribution. Westbury Formation (Moore 1861*b*; Vaughan 1904; Waters and Lawrence 1987).

'*Permophorus*' *angulatus* (Moore, 1861*b*)
Plate 15, figures 1–2

Description. Ovate-oblong, small, up to 20 mm long. Weak beaks near anterior. Three acute, slightly curved ribs extend from umbo to flattened, angular, postero-ventral margin.
Remarks. Moore's two figured specimens, a left valve M 150 (1861*b*, pl. 15, fig. 12; Pl. 15, fig. 1) and a right valve M 151 (pl. 15, fig. 13; Pl. 15, fig. 2) are quite dissimilar. The right valve is notably larger than the left and lacks the three acute ribs characteristic of the left. M 151 is too worn to be assigned definitely to this taxon.
Distribution. Appears to be confined to the Westbury Formation, rarely recorded.

Subfamily MYOCONCHINAE Newell, 1957
Genus MYOCONCHA J. de C. Sowerby, 1824
Myoconcha (*Myoconcha*) *psilonoti* Quenstedt, 1858
Plate 15, figure 3

Description. Medium-sized, modioliform, broad to somewhat spatulate, elongate. Long, narrow sub-marginal angulation bordered by furrow. Terminal beaks. Shell fairly thick but often not preserved.
Distribution. Rare, reported from the Langport Member in Somerset, but one specimen so allocated, BMNH L 77196, recorded as from the 'White Lias', Wurt Pit, Mendips, may be stratigraphically younger. Donovan (1958, p. 134) in fact concluded that the majority of fossils so far recovered from these secondarily chertified sediments were from lower Lias Group horizons. *Myoconcha psilonoti* has been recognized in the Langport Member in St Audries Bay, Somerset by one of us (PH). Also known from the Penarth Group at Aust Cliff, Gloucestershire.

Order MYTILOIDA Férussac, 1822
Superfamily MYTILOIDEA Rafinesque, 1815
Family MYTILIDAE Rafinesque, 1815
Subfamily MYTILINAE Rafinesque, 1815
Genus MYTILUS Linnaeus, 1758
Mytilus cloacinus Tutcher, 1908
Plate 15, figures 4–7

Description. Elongate, mytiliform, slightly arcuate. Outline acute anteriorly becoming broad and spatulate posteriorly with an evenly curved, smooth perimeter. Hingeline straight, about one-third length of shell. Beak terminal. A dorsally positioned carina, which commences near beak, broadens and becomes indistinct towards antero-ventral border. Com-

PLATE 15

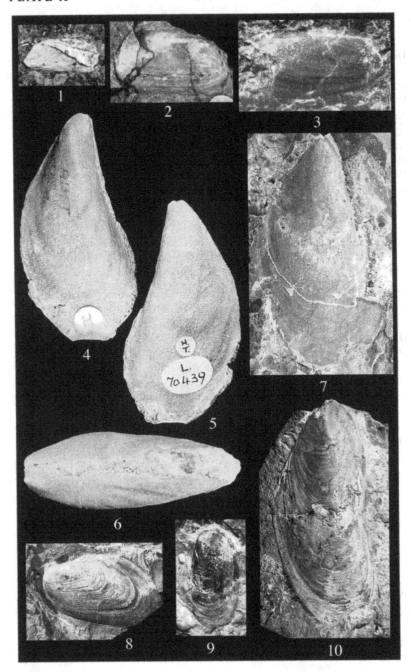

marginal growth halts are developed, but external ornament lacking except for fine growth lines. Often found as internal casts.

Remarks. The holotype of *M. cloacinus* (Tutcher 1908, unnumbered fig. opposite p. 20; herein Pl. 15, figs 4–6) is a nearly complete cast from the 'Ceratodus Bone Bed' of the basal Westbury Formation at Aust Cliff, now in The Natural History Museum (BMNH L 70439). Tutcher (1908) also found this species in bone beds at Sedbury (Monmouthshire) and Charlton (Bristol). Another specimen from the presumed same horizon as the holotype at Aust Cliff is illustrated as Plate 15, figure 7.

Distribution. Mostly found in the lowest beds of the Westbury Formation, especially in vertebrate-rich beds.

Subfamily MODIOLINAE Keen, 1958
Genus MODIOLUS Lamarck, 1799

Remarks. A number of species of this genus are recorded from the Penarth Group and from the base of the overlying Lias Group. James Sowerby described *Modiolus laevis* in 1812, *M. hillanus* and *M. minimus* in 1818. The provenance of the type specimens of these species is not well established, but all probably originate from horizons very low in the Lias Group. Numerous publications followed describing additional species from beds of similar age in Europe and these names were also applied to forms in British Lias Group and 'Rhaetic' strata by Moore (1861b) and others.

Richardson and Tutcher (1916) used the generic name '*Volsella*' Scopoli, 1777, when describing taxa from the Penarth and Lias groups. '*Volsella*' was ruled invalid (ICZN 1955 *in* Opinion 325) following upon a

---

EXPLANATION OF PLATE 15

RV=right valve, LV=left valve

Figs 1–2. '*Permophorus*' *angulatus* (Moore, 1861b); 'Flinty-bed', Westbury Formation; Beer Crocombe, Somerset. 1, syntype BRLSI M 150; external, LV; ×3. 2, syntype BRLSI M 151; worn, external, RV; ×2.

Fig. 3. *Myoconcha* (*Myoconcha*) *psilonoti* Quenstedt, 1858; NMW 93·93G·10; Penarth Group; Aust Cliff, Gloucestershire; external, RV; ×1·5.

Figs 4–7. *Mytilus cloacinus* Tutcher, 1908; 'Ceratodus Bone Bed', Westbury Formation; Aust Cliff, Gloucestershire. 4–6, holotype, BMNH L 70439; internal cast RV (4), LV (5) and dorsal view (6). 7, BRSUG 555/2; cast, RV. All ×1·5.

Figs 8–10. *Modiolus hillanus* (J. Sowerby, 1818). 8, BGS HIC 7515; Westbury Formation; Penarth Head, Glamorgan; LV; ×1·5. 9, BGS Bj 6039; Lilstock Formation; Stowell Park Borehole at 629·88 m, Northleach, Gloucestershire; external, RV; ×2. 10, SM J 58527; Westbury-on-Severn; RV; ×1·5.

primary ruling validating *Modiolus* Lamarck, 1799. Of the new taxa described by Richardson and Tutcher (1916) their *Volsella* (=*Modiolus*) *langportensis* was reported from both the Langport Member and basal Lias Group but is here regarded as a synonym of *Modiolus hillanoides* Chapuis and Dewalque, 1853, a name previously used for the larger species of *Modiolus* in the Penarth Group, for example by Ivimey-Cook (*in* Waters and Lawrence 1987). This taxon may also include the larger *Modiolus* material recorded by Moore (1861*b*, p. 505, pl. 15, fig. 26 only) as '*Modiola minima* Sowerby (=*Mytilus minuta* Goldfuss 1837)'. *Modiolus hillanoides* may in turn be a small or immature form of *Modiolus hillanus* (J. Sowerby, 1818). Certain smaller specimens, found rarely in the Penarth Group, are more 'bean' shaped (Moore, 1861*b*, pl. 15, fig. 27) and are here attributed to *M. minimus* (J. Sowerby, 1818). The earlier *Modiolus laevis* (J. Sowerby, 1812) has also been recorded, rarely, from the Langport Member, but these specimens may be juveniles of *M. hillanus*.

Modiolus hillanus (J. Sowerby, 1818)
Plate 15, figures 8–10

Description. Medium-sized, up to about 40 mm long, convex, sub-trapezoidal. Hinge line straight with sub-terminal umbones. Concave ventral margin, shell covered by fine commarginal growth lines. Shape rather variable from near parallel sided to markedly inflated and ovoid.

Remarks. James Sowerby's illustration (1818, pl. 212, fig. 2) is drawn with the body width slightly wider than that of the type specimen because of misleading residual rock material on the dorsal margin, which gives the appearance of greater body width. The narrower and less oblique form illustrated here (Pl. 15, fig. 9) is that often referred to *M. hillanoides*.

Distribution. Occurs sporadically in the Westbury and Lilstock formations. Recorded from the Westbury Formation near Watchet, Somerset (Dawkins 1864). Possibly present in the '*bristovi* limestones' at Lavernock (Ivimey-Cook 1974). Locally common in the silty marls and calcareous siltstones ('Watchet Beds') overlying the thin Langport Member limestones at Lavernock. Locally within the Langport Member elsewhere.

Modiolus minimus (J. Sowerby, 1818)
Plate 16, figures 1–2

Description. Small, up to about 10 mm long, elliptical, convex. Hinge line straight. Ventral margin nearly straight, anterior margin relatively pointed. Shell surface covered by fine commarginal growth lines.

Remarks. This species exhibits considerable variation both in degree of elongation and shell inflation. Some specimens attributed to this species may be juveniles of larger taxa.

Distribution. Rare, possible specimens in the Westbury Formation at Beer Crocombe, Moore (1861*b*, pl. 15, fig. 27 only), Westbury Formation near Watchet (Dawkins 1864). Cotham Member at Uphill, Somerset.

'*Modiolus*' *sodburiensis* Vaughan, 1904
Plate 16, figures 3–5

Description. Small, up to 20 mm long, sub-circular to ovoid. Extremely thin-shelled, juvenile stages strongly convex but adult stages flatter. Umbo close to anterior. Behind the beaks the hinge line appears straight and slightly oblique with broadest part of valve just past the end of hinge line, the anterior part evenly rounded, projecting slightly in front of beak. Posterior margin also evenly rounded. Juveniles oval, adults more rounded. External ornament of faint commarginal growth lines crossed by fine radial striae and grooves which impart weakness to shell which often breaks into nearly rectangular fragments. No pallial line, muscle impressions or hinge teeth known.

Remarks. The taxonomic affinities of this species are uncertain; it is almost certainly not a *Modiolus* and may have greater affinity with the genus *Bakevellia*. A similar form was figured by Quenstedt (1856, pl. 1, figs 12–13, 27, 33) but the generic attributions given there do not fit the features given above.

Distribution. Westbury Formation. Sporadic, may be only rarely recorded because the shell is very fragile and both the internal and external moulds may be very faint. Most frequently seen in unweathered mudstone cores from boreholes; however, Reynolds and Vaughan (1904, p. 196) found it in sandy micaceous beds in the Westbury Formation at Chipping Sodbury, Gloucestershire.

Subfamily LITHOPHAGINAE Adams and Adams, 1857
Genus LITHOPHAGA Röding, 1798?
Lithophaga? sp.
Plate 16, figure 6

Description. Small, sub-circular crypts or pits excavated into limestone surfaces. Crypts 2–15 mm in diameter, may extend to 100–120 mm depth into Langport Member limestones.

Remarks. These pits are reported, but not described, from limestones and limestone clasts, some at least of which appear to be penecontemporaneous with the enclosing beds although some, particularly from the Vallis Vale area, may not. Reported as common (Richardson 1911) in both the Cotham and Langport members. The range in size suggests the excavations may be attributable to more than one taxon and some at least may be better compared to the trace fossil ichnogenus *Gastrochaenolites*. For other Penarth Group trace fossils see Chapter 16.

Subclass LUCINATA Pojeta, 1978
Order LUCINOIDA Dall, 1889*a*
Superfamily LUCINOIDEA Fleming, 1828
Family LUCINIDAE Fleming, 1828
Subfamily MYRTEINAE Chavan, 1969
Genus MESOMILTHA Chavan, 1938?
Mesomiltha? sp.
Plate 16, figures 7–8

EXPLANATION OF PLATE 16

RV = right valve, LV = left valve

Figs 1–2. *Modiolus minimus* (J. Sowerby, 1818); two specimens on BGS Zf 1973; Lilstock Formation; Uphill road-cutting, Weston-super-Mare, Somerset. 1, external, RV. 2, external, RV. Both ×3.

Figs 3–5. '*Modiolus*' *sodburiensis* Vaughan, 1904; Westbury Formation. 3, syntype, BMNH L 25333; Lilliput, Chipping Sodbury, Gloucestershire; RV. 4, syntype, BMNH L 25334; locality as for 3; paired valves. 5, BGS Bj 6121; Stowell Park Borehole, Northleach, Gloucestershire at 633·1 m; external, RV. All ×2.

Fig. 6. *Lithophaga*? sp.; AS collection; top of Langport Member; Vallis Vale, Frome, Somerset; cross section of crypts; ×1.

Figs 7–8. *Mesomiltha*? sp. 7, BGS GSM 90659; Langport Member; near Lyme Regis; internal, RV. 8, BRSUG 531; Penarth Group; Borehole 46, Almondsbury, Bristol at 4·5 m; internal, ?RV. Both ×2.

Figs 9–10. *Lyriomyophoria postera* (Quenstedt, 1856); 'Flinty-bed', Westbury Formation; Beer Crocombe, Somerset. 9, BRLSI M 104; external, RV. 10, BRLSI M 105; external, LV. Both ×2.

Figs 11–12. *Tutcheria cloacina* (Quenstedt, 1856); Westbury Formation. 11, BGS BKD 0526; Chard Geothermal Borehole, Somerset at 131·04 m; internal, RV; ×1·5. 12, NMW 93·57G·31; Llanwern, Monmouthshire; external, RV; ×2.

Fig. 13. *Palaeocardita* cf. *austriaca* (von Hauer, 1853); BGS BDK 0771; Cotham Member; Twyning Borehole, Gloucestershire at 125·2 m; internal mould, LV; ×2.

Figs 14–15. *Astarte* sp. 14, BGS HIC 853; Langport Member, upper part of Bed 6 (Hallam 1960); Pinhay Bay, Devon; external mould, LV. 15, BGS GSM 31412; 'Rhaetic'; near Lyme Regis, Somerset; internal mould, LV. Both ×3.

Fig. 16. *Cardinia regularis* (Terquem, 1855); BGS AP 3521; Westbury Formation; Apley Barn Borehole, Witney, Oxfordshire at 179·6 m; external, RV; ×1·5.

PLATE 16

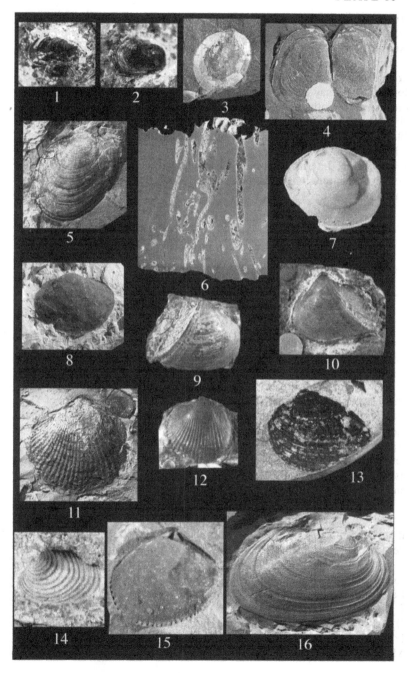

Description. Small, ovate-round, with fine, regular, equidistant commarginal ribs. Internal and external moulds of lucinid affinity.

Remarks. Bivalves of *Mesomiltha* aspect from the Penarth Group are inadequately known for proper appraisal, but similar round, often dark grey/black moulds occur in the Cotham Member in the Bristol area. These forms, which may compare to species of *Mesomiltha*, have been noted in one specific Cotham Member bed at Aust Cliff which also contains similarly preserved dark grey/black shell, fish and possibly insect remains. Plate 16, figure 8 shows a comparable form from a borehole at Almondsbury, Bristol.

Distribution. Internal and external moulds which may compare to this taxon occur rarely in the Lilstock Formation in Gloucestershire, Somerset and east Devon.

Subclass HETEROCONCHIA Hertwig, 1895
Order TRIGONIOIDA Dall, 1899
Superfamily TRIGONIOIDEA Lamarck, 1819
Family MYOPHORIIDAE Bronn, 1849
Genus LYRIOMYOPHORIA Kobayashi, 1954
Lyriomyophoria postera (Quenstedt, 1856)
Plate 16, figures 9–10

Description. Small, up to 15 mm long, rhomboidal, inflated. Prominent marginal carina, demarcating posterior area with conspicuous ornament of slightly curved, radial costae. Main body of shell smooth, ornamented with strong commarginal growth lines. Carina may appear denticulated where shell layers have suffered slight erosion. Left valve carina distinct and single, right valve may show a double carina separated by a longitudinal sulcus.

Remarks. *L. postera* occurs typically in thin shell beds associated with a few other taxa. The shell material is rarely preserved, consequently generic assignment is based on external form rather than knowledge of dentition and hingement. Both valves are usually present in similar quantities. Of all the Penarth Group bivalves *L. postera* is the one most commonly found with joined valves; some still articulated in closed position, others spread open as if still joined by the ligament. Some of these articulated specimens are also vertical to the bedding, indicating that *L. postera* was a shallow burrower.

Richardson (1911, p. 23), who rarely recorded *L. postera* or similar forms, referred some specimens to '*Myophoria emmerichi* Winkler, 1859'. Later authors considered this to be a synonym of the form now known as *L. postera*.

Distribution. Common, sometimes locally abundant, in the Westbury Formation and basal part of the Cotham Member. Rare in the Langport Member.

Order VENEROIDA Adams and Adams, 1856
Superfamily CARDITOIDEA Fleming, 1828
Family CARDITIDAE Fleming, 1828
Subfamily CARDITAMERINAE Chavan, 1969
Genus TUTCHERIA Cox, 1946
Tutcheria cloacina (Quenstedt, 1856)
Plate 16, figures 11–12

Description. Small, up to 22 mm long, sub-orbicular, inflated. Generally found as flattened specimens. Umbones slightly recurved. Entire surface bears variable number of quite broad, rounded or flat-topped radial ribs, which widen posteriorly and may be intersected by widely spaced commarginal growth lines. This intersection rarely makes the marked steps in the shell, with the production of a facet, commonly developed in later species of this genus. Inter-rib spaces reduce towards periphery, where they may be very narrow, appearing as a linear groove. Ribbing may be weak or absent near the umbo.

Remarks. Cox (1946) defined *Tutcheria* and gave details of the hinge structure deduced from species from the Lias Group; Will (1969) detailed ribbing and hinge structure, based on Quenstedt's original material. The stratigraphically earlier *T. cloacina* was described later (Cox 1949); it had previously been referred to the genera *Cardium*, *Cardita*, and *Palaeocardita* in Europe.

At Lavernock and some nearby localities both valves are common and are found in generally similar proportions with a significant percentage still articulated. The species displays considerable variation in the number and width of ribs, and in the degree of valve convexity. Vaughan (1904) was first to describe British examples of *T. cloacina*, only passing references having been made to it previously. Later it was widely recorded by Richardson (e.g. 1911).

Distribution. Westbury Formation and basal part of the Cotham Member.

Subfamily PALAEOCARDITINAE Chavan, 1969
Genus PALAEOCARDITA Conrad, 1867
Palaeocardita cf. *austriaca* (von Hauer, 1853)
Plate 16, figure 13

Description. Length about 13 mm. Equivalve, inequilateral, enlarged posteriorly. Numerous flat-topped ribs, which radiate from the umbo, widen towards margin and break up into granules. Beaks anterior. Hinge with well developed cardinals, distant posterior lateral in left valve.

Remarks. *Palaeocardita austriacum* is the type species of the genus (von Hauer 1853, pp. 736–737). The only British specimen so far identified is an internal mould, thus the ribbing is only faintly seen.

Distribution. One specimen only which was recorded from the Cotham Member of the Twyning Borehole, Gloucestershire. It is reported as common in the uppermost Rhaetian of Belgium (Joly 1907, p. 27).

Superfamily CRASSATELLOIDEA Férussac, 1822
Family ASTARTIDAE d'Orbigny, 1844
Genus ASTARTE J. Sowerby, 1816
Astarte sp.
Plate 16, figures 14–15

Description. Small, elliptical but for a small, prominent beak. Regularly spaced commarginal ribs invariably present immediately around beak but may fade out laterally. Inner margin of valve may be denticulate. Hinge has well developed cardinals.

Remarks. Generally found as casts and external moulds in the 'White Lias' but these are generally too poorly preserved for specific identification.

Distribution. Langport Member, Pinhay Bay, Devon. Noted by Will (1969) in the 'Contorta Schichten', which is broadly equivalent to the Westbury Formation, in Germany.

Family CARDINIIDAE von Zittel, 1881
Genus CARDINIA Agassiz, 1841
Cardinia regularis Terquem, 1855
Plate 16, figure 16

Description. Medium-sized, ovoid, up to about 45 mm long. Relatively compressed. Shell fairly thin. Commarginal growth lines regularly spaced. Lower border a smoothly convex arc. Curvature of both anterior and posterior borders assumes a similarly strongly convex arc. Beak located in anterior third of valve, acute but not very prominent. Lunule small.

Remarks. This taxon may be a junior synonym of *Cardinia ovalis* (Stutchbury, 1842), a form which occurs, rarely, in the *planorbis* Biozone but more abundantly in the overlying *liasicus* Biozone of the Hettangian in England and Wales (Palmer 1975).

Distribution. Throughout the Penarth Group, most abundantly towards the top of the Westbury Formation and the basal beds of the Cotham Member.

Superfamily CARDIOIDEA Lamarck, 1809
Family CARDIIDAE Lamarck, 1809
Subfamily PROTOCARDINAE Keen, 1951
Genus PROTOCARDIA Beyrich, 1845
Protocardia rhaetica (Merian, 1852)
Plate 17, figures 1–2

Description. Small–medium sized, up to 30 mm long, thin shelled, equilateral. Umbo slightly recurved. Posterior and anterior margins rounded. Part of posterior flank ornamented with 10–15 conspicuous regular riblets which pass into the main area of the shell which is relatively smooth, covered by inconspicuous fine commarginal growth lines. No carina is normally seen.

Remarks. This species is distinguished from the nearly contemporaneous *Protocardia philippianum* (Dunker, 1847) by the sculpture and form of the posterior ribbed area. In *P. rhaetica* the 10–15 ribs are in the post-carinal area but the carina itself is generally obsolete. In contrast, *P. philippianum* possesses five to seven ribs, restricted to a well-marked post-carinal area which may be set at an angle to the rest of the shell surface. Additionally, *P. rhaetica* has less prominent umbones, flatter valves and a relatively rounded posterior margin. Will (1969) suggested that morphological differences exist which would indicate the presence of several subspecies of *P. rhaetica* in Germany.

Distribution. Throughout the Penarth Group, most abundantly towards the top of the Westbury Formation and at the base of the Cotham Member. It overlaps with *P. philippianum* in lower Lias Group deposits.

Superfamily ARCTICOIDEA Newton, 1891
Family ARCTICIDAE Newton, 1891
Genus ISOCYPRINA Röder, 1882

Remarks. This genus is the current repository of several species of small, more or less ovate bivalves which have been allocated over the past two centuries to a variety of genera including *Axinus, Isocardia, Isodonta, Mesodesma, Myophoria?, Opis, Protocardia, Pullastra, Schizodus, Taeniodon* and *Tellinites*. In the Penarth Group specimens are generally poorly preserved and little is known about the hinge structure of most of the Rhaetian forms. Consequently their classification has been much influenced by overall shape as they generally show little ornament except for growth lines. They are often found as external or internal moulds and are usually compressed with much overlap of the valves. The classification of these forms is far from resolved; the following notes cover most of the widely quoted 'species' from the Penarth Group.

Subgenus EOTRAPEZIUM Douvillé, 1913

Remarks. The type species of *Eotrapezium* is *Mesodesma germari* Dunker, 1844. This subgenus is now used to accommodate Late Triassic–Early Jurassic members of the genus. These are principally 'suborbicular to elongate ovate; hinge with cardinal tooth 1 scarcely differentiated from A1, no 2a, lamina AIII-3a suppressed in most' (Cox, *in* Cox *et al.* 1969, p.

N648). These details of the hinge have not been observed in Penarth Group specimens, so the generic attribution of the species given below is based on their general external features. Moore (1861*b*) found a number of specimens, complete with calcite shells, in the 'Flinty-bed' at Beer Crocombe, upon which these descriptions are based, but he described no internal details.

Eotrapezium has also been considered a genus (Troedsson 1951), a practice often followed until recently (e.g. Ivimey-Cook 1974) and in post 1960 British Geological Survey publications.

Moore (1861*b*) recorded and described the following taxa, which he referred to the genus *Axinus* J. Sowerby, 1821, viz. *Axinus cloacinus* (Quenstedt, 1856), *A. concentricus* Moore, *A. depressus* Moore and *A. elongatus* Moore. Here these are regarded respectively as *Isocyprina* (*Eotrapezium*) *ewaldi, I. (E.) concentricum, I. (E.) depressum* and *I. (E.) germari*. The section concludes with a note about '*Pullastra arenicola* Strickland', to which various bivalves have been assigned, including some species of *Eotrapezium*. This is a taxon the use of which we recommend is not perpetuated.

EXPLANATION OF PLATE 17

RV=right valve, LV=left valve

Figs 1–2. *Protocardia rhaetica* (Merian, 1852); Westbury Formation. 1, BRLSI M 154; 'Flinty-bed'; Beer Crocombe, Somerset; external, LV; x 2. 2, BGS Bj 6070; Stowell Park Borehole, Northleach, Gloucestershire at 631·16 m; external, valve pair; ×1·5.

Figs 3–4. *Isocyprina* (*Eotrapezium*) *ewaldi* (Bornemann, 1854); Westbury Formation. 3, BGS GSM 54071; Beer Crocombe, Somerset; LV. 4, BRLSI M 95; 'Flinty-bed'; Beer Crocombe, Somerset; LV. Both ×3.

Figs 5–8. *Isocyprina* (*Eotrapezium*) *concentricum* (Moore, 1861*b*). 5–7, 'Flinty-bed'; Westbury Formation; Beer Crocombe, Somerset. 5, syntype, BRLSI M 100; RV. 6, syntype, BRLSI M 99; LV. 7, BRLSI M 98, 'variety of *A. concentricum*'; LV. 8, BGS Bl 7986; Westbury Formation; Upton Borehole, Burford, Oxfordshire at 174·17 m; articulated valves. All×3.

Fig. 9. *Isocyprina* (*Eotrapezium*) *depressum* (Moore, 1861*b*); holotype BRLSI M 101; 'Flinty-bed', Westbury Formation; Beer Crocombe, Somerset; external, RV; ×2.

Fig. 10. *Isocyprina* (*Eotrapezium*) *germari* (Dunker, 1844); BRLSI M 102; 'Flinty-bed', Westbury Formation; Beer Crocombe, Somerset; external, LV; ×3.

Figs 11–13. *Pholadomya* sp.; Langport Member, 'White Lias', 'Rhaetic'. BMNH L 77177; Rockhill, Radstock, Somerset. 11, LV. 12, RV. 13, BMNH L 29450; 11–12, Windmill Hill, Walton, Street, Somerset; LV. All ×1·5.

PLATE 17

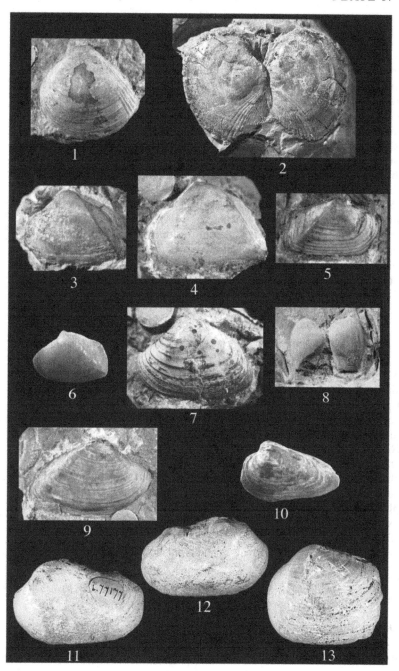

Isocyprina (Eotrapezium) ewaldi (Bornemann, 1854)
Plate 17, figures 3–4

Description. Shell rather triangular, smooth, with fine commarginal growth lines. Equivalve, inequilateral, fairly convex, umbones slightly anterior. Anterior rounded, posterior slightly extended, with a depressed ridge extending from umbo to postero-ventral margin. This defines an arcuate bounded small area in which growth lines curve back towards the umbo.

Remarks. Moore (1861*b*) regarded the specimens that he attributed to *Axinus cloacinus*, here attributed to *I. (E.) ewaldi*, as the most common form in the 'Flinty-bed', but overall *I. (E.) concentricum* may be more widespread in the Penarth Group. Will (1969) recorded specimens up to 15 mm long and 12·3 mm high. The principal difference between this form and *I. (E.) concentricum* is that the arcuate posterior area has recurved growth lines, rather than the squared-off area with growth lines ending almost at right angles to the dorsal margin of *I. (E.) concentricum*.

Distribution. Common in the Westbury Formation; also occurs in the Lilstock Formation as in the Langport Member at Pinhay Bay, Devon (Hallam 1960).

Isocyprina (Eotrapezium) concentricum (Moore, 1861*b*)
Plate 17, figures 5–8

Description. Small, up to 12 mm long, triangular to elongate, compressed. Anterior margin short, rounded; posterior margin extended, straight and sloping, bounded by a carina from umbo to postero-ventral margin. Umbones somewhat raised, convex and anterior of mid-line. Surface covered by fine commarginal growth lines which cross posterior area to meet postero-dorsal margin at about 90°.

Remarks. There is some variation in form with some specimens more arcuate posteriorly and approaching *Isocyprina (Eotrapezium) ewaldi* in shape, so records of the species may be confused. Moore (1861*b*) considered this form comparable to ones illustrated by Quenstedt (1856, pl. 1, figs 28, 30). Richardson (1905, 1911) almost certainly encountered examples of this species but generally referred them to *Schizodus ewaldi* (Bornemann).

Distribution. Westbury Formation and basal Cotham Member.

Isocyprina (Eotrapezium) depressum (Moore, 1861*b*)
Plate 17, figure 9

Description. Small, elongate–ovoid in shape. Equivalve, slightly inequilateral. Shell rather flat, evenly inflated, with a posterior carina delimiting a restricted dorsal area. Umbo raised, slightly anterior to submedian, umbonal angle apparently around 110°. Commarginal ribbing apparently meets dorsal margin at high angle. Details of hingeline and internal morphology unknown. Holotype (BRLSI M 101) is 18 mm long and 10 mm high dorso-ventrally.

Remarks. The holotype is preserved in limestone and may not be completely exposed. The authors are unaware of other specimens. Holotype is more depressed than *I. (E.) ewaldi*, with the edge of the shell apparently closer to the carina, which resembles a fold rather than a ridge. Moore (1861*b*, p. 503) gave no indication of particular rarity, but only Etheridge (1872) has cited this species since, from Penarth. There remains a possibility that *I. (E.) depressum* is founded on a slightly aberrant specimen of *I. (E.) concentricum*, and more material is clearly needed for a meaningful assessment.

Distribution. Lower part of the Westbury Formation.

Isocyprina (Eotrapezium) germari (Dunker, 1844)
Plate 17, figure 10

Description. Small, up to 16 mm long, sub-orbicular to elongate oval, inequilateral, relatively compressed. Anterior and posterior margins rounded although posterior margin for a short distance may appear relatively straight and at right angles to hinge line. Umbones relatively prominent about one-third from anterior end, only slightly raised. Surface covered in regular commarginal growth lines. Posterior carina weak or obsolete.

Remarks. The name *Isocyprina (Eotrapezium) germari* has been used for a form common in the lower beds of the Westbury Formation e.g. Ivimey-Cook (*in* Waters and Lawrence 1987). Moore's species *Axinus elongatus* (1861*b*, p. 503, pl. 15, fig. 18) was founded on two specimens from the 'Flinty-bed' at Beer Crocombe; it is here considered to be a junior synonym of Dunker's taxon (1846, p. 40, pl. 6, figs 20–22). *Axinus elongatus* was recorded by Etheridge (1872), but otherwise has rarely been mentioned since its first description. The shell illustrated here is Moore's figured specimen of *A. elongatus* as this is one of few specimens from Britain with the shell preserved.

Distribution. Lower part of the Westbury Formation. Troedsson (1951) recorded this form from the lower part of the Helsingborg Stage in Sweden (probably equivalent to the *planorbis* Zone of the Hettangian) whilst Dunker (1846) recorded it from beds of Rhaetian age in Germany.

'*Pullastra arenicola*' Strickland, 1843

Description. 'Nearly a perfect oval, depressed, nearly smooth but with faint concentric striation towards the margins. The apex is about half way between the middle of the shell and the anterior end. The general outline closely resembles that of the recent *Pullastra aurea* of Britain. Max. length 7 lines, breadth 4·5 lines, but ordinary size is less' (Strickland 1843, p. 17, based on casts and moulds).

Remarks. This name was applied to many specimens from the 'Rhaetic', particularly in the nineteenth century. However, it has never been adequately described or figured and is best rejected as a taxon.

Phillips (1871, p. 107, pl. 7, figs 6–12) described and figured specimens that he attributed to this taxon. The figures were indifferently drawn and do not aid interpretation, especially as they do not fit the original description. The genus *Pullastra* G. B. Sowerby, 1826 is now regarded as a synonym of the venerid genus *Venerupis* Lamarck, 1818, of Tertiary–Recent age (Keen, *in* Cox *et al*. 1969, p. N685).

The name has been applied to generally ovate casts and moulds, often in fine sandstone or siltstone lenticles which occur in the lower part of the Westbury Formation in south-west Britain. The casts and moulds are probably principally of *Isocyprina* (*Eotrapezium*) spp. but may include other taxa.

Subclass ANOMALODESMATA Dall, 1889*a*
Order PHOLADOMYOIDA Newell, 1965
Superfamily PHOLADOMYOIDEA Gray, 1847
Family PHOLADOMYIDAE Gray, 1847
Genus PHOLADOMYA G. B. Sowerby, 1823
Pholadomya sp.
Plate 17, figures 11–13

Description. Medium-sized, strongly inequilateral, equivalve myid. Ovate to sub-trigonal. Umbones broadly rounded, variably prominent, more or less anterior. Valves gape to posterior, anterior gape narrow if present. Pustulation very delicate. Ornament may be strong or weak, with well-developed radial ribs or ridges; some may bear tubercles. Commarginal rugae. Pallial sinus broad. (Based on Cox, *in* Cox *et al*. 1969, p. N827). The poorly preserved material available is small- to medium-sized, possibly juvenile, rarely reaches 30 mm long and has indistinct ornament.

Distribution. Richardson (1911, p. 41) cited *Pholadomya* (of *P. glabra* group?) from the Langport Member at Charlton Mackrell, Somerset. *Pholadomya glabra* was described by Agassiz (1842, p. 69, figs 12–14). The specimens figured here are from near Radstock and Street, Somerset. Dawkins (1864, p. 407) referred specimens from the 'White Lias' to this genus but these are probably from a higher horizon as they were recorded in association with *Pinna* at Saltford, Gloucestershire. Whilst specimens of this genus are rare and poorly known in Britain the genus is better known from the 'Contorta Schichten' and 'Zone of the Infra-Lias' in Europe.

Family CERATOMYIDAE Arkell, 1934
Genus PTEROMYA Moore, 1861*b*

Description. 'Subovate, sub-equilateral or inequilateral to a varying extent, not strongly inflated, slightly to moderately inequivalve, right valve the more gibbose and with its umbo the more elevated; valve margins not gaping; umbones broad, protruding very little; postero-dorsal and posterior margins forming continuous, strongly convex curve or else meeting in obtuse angle, in which case very obtuse posterior ridge may be present in one or both valves; hinge structure as in *Ceratomya* and *Gresslya*; adductor scars and pallial line not yet observed, the latter probably without a sinus; surface with concentric undulations or ridges, or merely with coarse growth threads' (Cox, *in* Cox *et al.* 1969, p. N841).

Remarks. Cox (1963) studied transverse sections of the hinge line of *Pteromya*, which satisfied him that this genus could be distinguished from *Pleuromya*. The marginal undulations, folds and ridges on the outer shell may be restricted to earlier growth stages. The shell is usually thin with some growth lines which may be well marked.

Pteromya crowcombeia Moore, 1861*b*
Plate 18, figures 1–2

Description. Elongate, oval, inequilateral, myid-like shells up to about 26 mm long. Left valve has obtuse ridge from umbo to postero-ventral corner of shell, separating flattened postero-ventral area by well defined obtuse outer angle from flank of shell. In the contrasting right valve, refigured by Cox (1963, pl. 79, fig. 2), postero-dorsal and posterior margins usually form strongly convex, uninterrupted curve terminating at sharp angle at posterior end of ventral margin with no trace of posterior ridge; however, some specimens have developed a faint posterior ridge. Right valve consistently more strongly convex than left. Both valves variable in both shape and ornament of irregular commarginal folds and ridges.

Remarks. Cox (1963) redescribed the species in detail and selected as lectotype, from nine available syntypes, a left valve, M 136 from the Moore Collection in Bath (Cox 1963, pl. 79, fig. 1). He also placed the small specimen described by Moore (1861*b*, p. 506, pl. 15, fig. 24) as *Pteromya simplex*, in synonymy with *P. crowcombeia*.

Distribution. Westbury Formation in the south-west of England. Rarely recorded from other parts of Europe. A specimen described and figured as *Pteromya* sp. nov.? (Melville 1956, pp. 76, 94, pl. 6, fig. 15) was, in Cox's (1963) view, very similar to this species. However, it was recovered from the *laqueus* Biozone of the Lias Group in the Stowell Park Borehole, Gloucestershire.

Pteromya aff. *crowcombeia* Moore, 1861*b*
Plate 18, figures 3–5

Description. Elongate ovoid, beak fairly prominent. Both valves similar in convexity, reaching 22 mm long; dorsal and posterior margins meet at well marked obtuse angle. Traces of weak posterior ridge in some specimens. No details available on nature of hinge, pallial line or muscle scars.

Remarks. In size and proportions Cox (1963) considered that these specimens are similar to *P. crowcombeia* rather than the near age equivalent *P. langportensis* (Richardson and Tutcher) (see below), which is larger and less elongate. Cox (1963) found no specimens where the dorsal and posterior margins form a single continuous curve as in the specimens of *P. crowcombeia* from the Westbury Formation. Specimens occur as internal moulds of both valves scattered on a single bedding plane at Garden Cliff, Westbury-on-Severn. Cox (1963) was unable to confirm the identity of further examples attributed to this taxon from Yorkshire (Blake 1872; Tate 1876).

Distribution. Only known from 'Bed 17' of Etheridge (1865, p. 238) at Garden Cliff. Cox (1963) attributed this bed to the 'Upper Rhaetic' Cotham Beds. It forms the lower part of a 120 mm thick limestone, the upper leaf being the Pseudomonotis Bed (Etheridge 1872). The restricted fauna of this bed is not age diagnostic and occurs in a thickness of shales,

---

EXPLANATION OF PLATE 18

RV=right valve, LV=left valve, cop.=copy of

Figs 1–2. *Pteromya crowcombeia* Moore, 1861*b*; 'Flinty-bed', Westbury Formation; Beer Crocombe, Somerset. 1, syntype BRLSI M 136, lectotype; external, LV. 2, syntype BRLSI M 138a, paralectotype; external, RV. Both ×2.

Figs 3–5. *Pteromya* aff. *crowcombeia* Moore, 1861*b*; Cotham Member Bed 17; Garden Cliff, Westbury-on-Severn, Gloucestershire. 3, BMNH L 67461; LV; ×2. 4, BMNH L 67461; whole slab; cop. Cox 1963, pl. 79, fig. 6; ×1. 5, BMNH L 38237; cop. Cox 1963, pl. 79, fig. 7; ×1.

Figs 6–9. *Pteromya langportensis* (Richardson and Tutcher, 1916); Langport Member. 6–8, lectotype, BMNH L 70446; Radstock Grove, Radstock, Somerset. 6, LV. 7, RV. 8, dorsal view, RV uppermost. 9, BMNH L 30975; Rockhill, Radstock, Somerset; dorsal view with valve twisted to opposite side. All ×1.

Figs 10–11. *Pleuromya*? sp.; 'Flinty-bed', Westbury Formation; Beer Crocombe, Somerset. 10, BRLSI M 131; RV. 11, BRLSI M 141; RV. Both ×1.

Figs 12–13. *Cercomya praecursor* Quenstedt, 1856; Westbury Formation. 12, BGS GK 1203; stream 300 m north-north-east of Queen Charlton Church, Keynsham, Bristol; RV. 13, BGS GSM 7826, holotype of *C. praecursor* var. *pylensis* Tawney; Quarella Sandstone; Pyle, near Bridgend, Glamorgan; RV. Both ×1.

PLATE 18

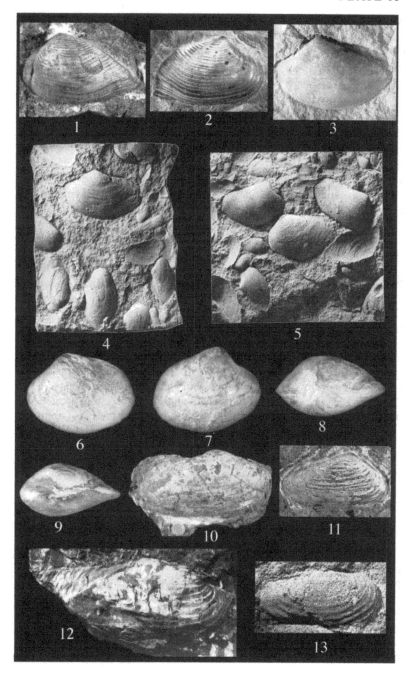

which may equate with some part of the Langport Member, but which lithologically might also be correlated with the basal part of the Lias Group (Warrington and Ivimey-Cook 1992).

Pteromya langportensis (Richardson and Tutcher, 1916)
Plate 18, figures 6–9

Description. Moderately equivalve, high but not very elongate, length commonly up to 30 mm but may reach 40 mm. Right valve more strongly inflated and with higher umbo than left. Internal moulds of conjoined valves show a continuous sharp ridge around the complete valve, precluding a gape. Valves meet without overlap. Strongly asymmetrical convex ventral margin. No posterior ridge on internal moulds of either valve, nor distinct postero-dorsal angle to valves. Valve margin not always in a single plane as, whilst most specimens show the posterior end of shell bending to the right (Pl. 18, fig. 8), in some the shell is bent to the left (Pl. 18, fig. 9).

Remarks. Cox (1963) designated and figured one of Richardson and Tutcher's moulds (BMNH L 70446) as lectotype for this species. Richardson and Tutcher (1916, p. 52, pl. 8, fig. 5) originally described their material as a variety of their new species *Pleuromya tatei*. They placed this new species in *Pleuromya* as the two valves were similar, unlike the dissimilar valves of *Pteromya* which are one of its principal characters. Cox (1963) transferred *P. tatei* to *Pteromya* and raised the variety *langportensis* to species rank. Whilst *P. langportensis* occurs in the Langport Member, *P. tatei* is not convincingly recorded from these beds, but is abundant in the succeeding Preplanorbis Beds of the Lias Group.

Distribution. Common in the Langport Member of the Radstock area, Langport and some other parts of Somerset.

Family PLEUROMYIDAE Dall, 1900
Genus PLEUROMYA Agassiz, 1842
Pleuromya? sp.
Plate 18, figures 10–11

Description. Equivalve, oval to triangular, moderately inflated with posterior gape. Umbones generally anterior of midline, rounded but not protruding strongly. No demarcated areas, but regular concentric ribbing often present. Radial rows of minute pustules may be present.

Remarks. This genus may be appropriate for the specimens previously referred by Moore (1861*b*) to *Myacites* Schlotheim. Plate 18, figure 10 shows the holotype of *Myacites striato-granulata* Moore (1861*b*, p. 506, pl. 16, fig. 1). This and a further specimen in The Natural History Museum

(BMNH L 81953 ex Moore Collection), are recorded from the 'Flinty-bed' at Beer Crocombe. The apparently rather triangular bivalve recorded as '*Anatina? praecursor*' Moore (1861*b*, pl. 16, fig. 3; Pl. 18, fig. 11) *non* Quenstedt, 1856 may also be referable to this genus.

Distribution. A number of impressions and moulds probably referable to *Pleuromya* have been found in the Langport Member but are usually poorly preserved and often their exact stratigraphical horizon is unclear. Moore's (1861*b*) specimens are reputedly from the Westbury Formation.

Superfamily PANDOROIDEA Rafinesque, 1815
Family LATERNULIDAE Hedley, 1918
Genus CERCOMYA Agassiz, 1843
Cercomya praecursor Quenstedt, 1856
Plate 18, figures 12–13

Description. Equivalve, elongate, moderately inflated especially at anterior, less inflated towards the lower, tapering posterior end which may become slightly upcurved. Umbones sub-central, depressed to generally level with postero-dorsal margin. Flanks have six to ten rounded concentric folds. Weak ridge extends from umbones to postero-ventral margin. Surface weakly pustulate. Around 35 mm long and 18 mm wide.

Remarks. The taxon described and figured by Quenstedt (1856, pl. 1, fig. 15) showed the tapering posterior end bounded dorsally by a strongly arcuate curve. Specimens of this morphology occur but others are more rectangular. This possibly accounts for the difficulty in attribution discussed by Moore (1861*b*, p. 507) and why his plate 16, figure 2 is recorded as '*Anatina? suessi* Oppel' but resembles *C. praecursor*. This form, of which no internal details have been described, may be better regarded as belonging with the Pholadomyidae, perhaps in *Pachymya* (*Arcomya*) or *Homomya*. The genus *Anatina* Lamarck, 1818 is now regarded as a junior synonym of *Laternula* Röding, 1798 (Cox *et al.* 1969) from which the family takes its name. Tawney (1866, p. 88, pl. 4, fig. 5; Pl. 18, fig. 13) described a variety (*C. praecursor* var. *pylensis*) from sandstones in Mid-Glamorgan; this is shorter posterially and has stronger marked folds around the posterior end.

Distribution. Uncommon, found in the Westbury Formation, especially in calcareous and sandy horizons, where the distinctive shape is easily recognizable.

9. ARTHROPODS 1: CRUSTACEANS

by IAN D. BOOMER, CHRISTOPHER J. DUFFIN *and* ANDREW SWIFT

Large arthropod fossils are very rare in the Penarth Group. The majority of recorded forms are microscopic, such as the ostracods, or very small e.g. the concostracans. At certain levels these forms may be abundant and dominate the fauna. Descriptions of larger, but still relatively small, arthropods are restricted to two decapod crustaceans, a possible isopod and a cirripede; a few other taxa have been reported but remain undescribed (Duffin 1978a; Förster and Crane 1984). However, even these latter 'large' forms are found at only a handful of localities and are incomplete. Only the trace fossil record gives any indication of the actual significance of large arthropods in the overall Penarth Group biota, with many traces being attributable to members of this phylum (see Chapter 16).

On Plate 19, GSM Mik (J) specimens are held in the collections of the British Geological Survey, Keyworth, the TCD specimen is held in the collection of Trinity College, Dublin and the remaining two specimens are held in the A. Swift collection, Department of Geology, University of Leicester. On Plate 20, figures 1–4 are held in the A. Swift collection, Department of Geology, University of Leicester (prefix BC), figures 5–8 in the collections of The Natural History Museum and figure 9 in the Moore Collection at the Bath Royal Literary and Scientific Institute (BRLSI).

Phylum ARTHROPODA Siebold and Stannius, 1845
Subphylum CRUSTACEA Pennant, 1777
Class OSTRACODA Latreille, 1806

Remarks. Ostracods are small crustaceans (generally 0·3–1·5 mm long) characterized by a dorsally-hinged bivalved carapace enclosing the body and limbs, from which protrude appendages for locomotion, feeding and reproduction. They inhabit a wide range of marine and non-marine environments and have a fossil record from at least the Ordovician to the Recent. With the ostracods, as with many fossil groups which occur in the latest Triassic, it is often difficult to distinguish true evolutionary events from colonization events during the regime of global sea level rise at that time.

Latest Triassic ostracod assemblages have been described from the European Alpine region (Bolz 1971a, 1971b; Kristan-Tollmann 1971), the Middle East (Kristan-Tollmann *et al.* 1980), the Far East (Gramm 1975) and South-East Asia (Kristan-Tollmann and Hasibuan 1990). Ostracod assemblages described from these areas are considered to be exclusively marine, whereas the collections from the Penarth Group, as well as containing fully marine taxa, also include several marginal marine/ lagoonal through to brackish/freshwater taxa, especially in the Cotham Member.

The earliest works on British Triassic ostracods were those of Duff (1842), Brodie (1845) and Jones (1894). The last publication was the most important source of taxonomic and other information for Penarth Group ostracods for more than 80 years. Anderson (1964) was first to consider Penarth Group ostracods in detail. He described assemblages from exposures and boreholes in south-west Britain, South Wales and the English Midlands, and compared the stratigraphical ranges of British species with records from North-West Europe. However, the taxonomic status of some species erected by Anderson (1964) remains to be confirmed. He figured only line drawings and erected certain new genera on the basis of single specimens; thus some of the work is insecure. Bate (1978a) undertook a stratigraphical review of British Triassic ostracods. With the exception of ostracods recorded from Anisian horizons in the Withycombe Borehole of Oxfordshire (Bate 1978b), Penarth Group ostracods of Rhaetian age are the only representatives of the group recorded from British Triassic rocks.

Ostracod assemblages from the Penarth Group are generally sparse and of low diversity, although high abundance levels do occur. The best known collections have been recovered from coastal sections along the Bristol Channel. The stratigraphical distribution of ostracods in these sections was described by Lord and Boomer (1990, see also for range charts). Ainsworth (1989a, 1989b, 1990) and Ainsworth *et al.* (1989) recorded ostracods from rocks recovered from exploration wells off western Ireland which may be broadly equivalent to some part of the onshore Penarth Group sequence, but detailed stratigraphical resolution is lacking for these cores. The stratigraphical review undertaken by Bate (1978a) has been revised (Ainsworth and Boomer in press).

British Triassic ostracods generally belong to smooth unornamented or weakly ornamented genera and are often preserved as carapaces. Thus, there are few external features to aid specific (and in some cases generic) identification, while the closed carapaces deny access to the often taxonomically significant internal features such as the muscle scars and hingement. Nevertheless, it is possible to assess the evolutionary position of many Penarth Group ostracods in relation to the marine assemblages from the overlying Lias Group.

Some Penarth Group ostracod assemblages have been attributed to non-marine, brackish or marginal marine/lagoonal environments. The non-marine taxa are represented by species of *Darwinula*, an extant genus which can be found today in rivers, freshwater lakes and the upper reaches of estuaries throughout the world. Ostracods of the genus *Lutkevichinella* from the Penarth Group are also considered here to be freshwater forms.

This latter genus is thought to be a limnocytherid and resembles the extant genus *Limnocythere*, which occurs in similar environments to *Darwinula*.

Supposed brackish water ostracod assemblages from the Penarth Group generally comprise a mixture of freshwater and more marine taxa. It is important in such cases to consider the taphonomy of the assemblage. A mixture of taxa suggests a number of possibilities: (1) some taxa may be living beyond their 'normal' salinity range; (2) there may have been reworking of older sediments; or (3) the mixing may be contemporaneous e.g. storm surges may have carried marine taxa into freshwater areas or floods may have washed freshwater forms into the sea. A fourth possibility is that the environment in which the ostracods were living was at the interface of marine and freshwater systems, as in an estuary, where mixing of post-mortem assemblages would be expected.

One Penarth Group genus, *Rhombocythere*, is known only from the Triassic and its ecology has been inferred indirectly from its observed association with species of the generally freshwater genus *Darwinula*. It did not survive into the marine conditions of the Early Jurassic and it is concluded that it too was restricted to freshwater and/or possibly brackish water environments.

Marine assemblages from the Penarth Group generally comprise species which survive into, or have descendants in, the Lias Group. Some of these marine species may have been capable of withstanding reduced salinity conditions, as they have been found in association with *in situ* freshwater species. It is also probable that some 'marine' taxa may have adapted to hypersaline lagoonal conditions, which may explain some of the high-abundance monospecific floods of metacopine species observed at highest Penarth Group and lowest Lias Group levels. Metacopina are unornamented marine ostracods with ovate, inflated carapaces and often dominate assemblages in the Lias Group until their final global extinction in the earliest Toarcian. The metacopine genus *Ogmoconchella* is common in the Penarth Group, dominating assemblages in the higher beds.

There are two other main groups of marine ostracods in the Penarth Group: Cytheroidea, represented by *Ektyphocythere*; and Platycopina, represented by *Cytherella* and *Cytherelloidea*. There is little evidence to suggest that these genera inhabited anything other than shelf sea environments.

SPECIES DESCRIPTIONS

Order PODOCOPIDA Müller, 1894
Suborder METACOPINA Sylvester-Bradley, 1961
Superfamily HEALDIOIDEA Harlton, 1933
Family HEALDIIDAE Harlton, 1933
Genus OGMOCONCHELLA Gründel, 1964

Description. Left valve larger than right. Left posterior margin more tumid, left valve usually higher and more broadly rounded in lateral view, than right.

Remarks. As with *Darwinula* this genus is often difficult to identify to species level due to the lack of distinctive external features. The main diagnostic characters are the outline in lateral and dorsal view, the degree of overlap and the presence and development of anterior flanges and posterior spines. Two species are described here; Anderson (1964) recorded a further four allied species from the Penarth Group, which are also referable to *Ogmoconchella*, but which Anderson (1964) considered belonged to the genus *Hungarella* Mehés, 1911: *H. moorei* (Jones, 1894), *H. elongata* (Blake, 1876), *H. owthorpensis* Anderson, 1964 and *H. caudata* Anderson, 1964. The taxonomic status of these four species remains uncertain.

Ogmoconchella martini (Anderson, 1964)
Plate 19, figure 1

Description. Ovate carapace characterized by short, blunt caudal process around mid-height on posterior margin. Well-developed 'fingerprint' ornament of fine ribbing on lateral surfaces.

Remarks. The fine ribbing on the lateral surfaces, only weakly developed on other *Ogmoconchella* species, is well seen in this species.

Distribution. Recorded from the Westbury Formation at two localities: Hampstead Farm Quarry, Chipping Sodbury, Gloucestershire (in association with *Ektyphocythere cookiana*), and the Plattlane Borehole, Whixall, Shropshire (Anderson 1964; Boomer 1991*b*). Also very common in the Langport Member.

Ogmoconchella bristolensis (Anderson, 1964)
Plate 19, figure 2

Description. Carapace ovoid in lateral outline, quite tumid. Left valve bears small anterior flange, right valve has anterior lip. Lateral surfaces very faintly ribbed.

Remarks. Less robust than *O. martini* and *O. owthorpensis*, with a more elongate outline and more acute dorsal margin in lateral view.

Distribution. Fairly common in the Langport Member.

Suborder PLATYCOPINA Sars, 1866
Superfamily KLOEDENELLOIDEA Ulrich and Bassler, 1908
Family CYTHERELLIDAE Sars, 1866
Genus CYTHERELLA Jones, 1894

Description. Carapace compressed. Valves smooth, lacking spines or marginal flanges. Right valve larger than left.

Remarks. When compared with Metacopina, the carapace is more compressed, with relative valve sizes reversed. *Cytherella* is considered to be exclusively marine; extant representatives inhabit shallow marine shelf seas and open oceans.

Cytherella plattensis Anderson, 1964
Plate 19, figure 3

Description. Carapace smooth, ovate in lateral view. In dorsal view anterior margin compressed; posterior margin blunt, almost square. Valves overlap to greatest extent along mid-anterodorsal and ventral margins.

Remarks. Figures 67 and 68 on plate 12 of Anderson (1964) are upside down.

Distribution. A common constituent of Langport Member assemblages, but also occurs less abundantly at lower levels in the Penarth Group.

Genus CYTHERELLOIDEA Alexander, 1933

Description. Carapace similar to *Cytherella* but differs most notably in possession of strong reticulation and/or ribbing.

Cytherelloidea cf. *pulchella* Apostolescu, 1959
Plate 19, figure 4

Description. Carapace small, ovate, sub-quadrate with strong (antero-ventral-postero-) marginal rib and three weaker transverse ribs. Lateral surfaces bear fine regular reticulation. Anterior margin broadly rounded, posterior margin bluntly pointed with extremity at about mid-height.

Remarks. Probably not synonymous with Apostolescu's species but certainly closely related to it. The relatively elaborate ornamentation makes this a distinctive species in Penarth Group collections.

Distribution. Common in the Langport Member, occasionally found in lower beds.

Suborder PODOCOPINA Sars, 1866
Superfamily BAIRDIOIDEA Sars, 1888
Family BAIRDIIDAE Sars, 1888
Genus BAIRDIA McCoy, 1844

Description. Anterior broadly rounded, posterior narrow to acuminate, lateral surfaces typically smooth. Dorsal margin generally comprises anterior, central and posterior straight to slightly convex elements.

Remarks. *Bairdia* has a fossil record from the Palaeozoic to Recent and generally indicates fully marine conditions. The few recorded Penarth Group specimens are mostly juveniles, and are also broken.

Bairdia sp.
Plate 19, figure 5

Description. Carapace slender, elongate posteriorly, concave antero- and postero-dorsally (Bate 1978*a*).
Remarks. It is often difficult to identify juvenile ostracod specimens to specific level, particularly when the specimens lack distinctive ornament or are broken as in the case of the illustrated specimen (Pl. 19, fig. 5), which comes from the Cotham Member at Long Itchington Quarry, Warwickshire. It is probably conspecific with the adult specimen of *B. anisica* Kozur figured by Bate (1978*a*) from a locality at nearby Combrook, Warwickshire. However, the holotype of *B. anisica* is more elongate in lateral outline than the species recovered from the Cotham Member in Britain. Anderson (1964) originally referred the Combrook specimen to cf. *B. molesta* Apostolescu, a Liassic species, which also has a different lateral outline (see Bate 1978*a*). Modern species of the bairdiid group are exclusively marine.
Distribution. A very rare ostracod in the Cotham Member. Recorded only from Long Itchington Quarry and from an old railway cutting near Combrook, both in Warwickshire. Recorded from the Langport Member and the *angulata* Zone of the Lower Jurassic at Lavernock by Lord and Boomer (1990).

Superfamily CYTHEROIDEA Baird, 1850
Family LIMNOCYTHERIDAE Klie, 1938
Genus LUTKEVICHINELLA Schneider, 1956

Description. Carapace medium to large, generally sub-rectangular. Dorsal margin straight, ventral margin straight to sinuous. Central sulcus present, valves smooth to strongly reticulate. Overlap very weak, left valve larger.
Remarks. Limnocytheridae are considered to be almost exclusive to freshwater environments and have a fossil record from Carboniferous to Recent. *Lutkevichinella* is excluded from the Metacopina because of its less well developed valve overlap and more quadrate lateral outline. Platycopina are more robust and possess a distinct right valve over left overlap. The genus is an important palaeoenvironmental indicator and is commonly recorded in Penarth Group assemblages together with *Darwinula*. In modern freshwater environments *Limnocythere* is often found in association with *Darwinula*.

Lutkevichinella fastigata Ainsworth, 1989*b*
Plate 19, figure 6

Description. Carapace small, sub-triangular. Anterior and posterior margins compressed, the former broadly rounded, the latter narrow. Greatest height in front fifth of valve. Mid-valve region bears shallow open reticulae.

Remarks. This species is from highest Triassic horizons of offshore west Ireland, where completely smooth congeneric species have also been recorded (e.g. *L. hortonae* Ainsworth, 1989*b*). Correlation with the Penarth Group onshore is uncertain and there are no records from onshore localities. Another smooth species represented by a few poorly preserved carapaces has been recorded from the top of the Cotham Member at Long Itchington Quarry, Warwickshire.

Distribution. Recorded only from highest Triassic horizons offshore west Ireland.

Family LIMNOCYTHERIDAE Klie, 1938
Genus ALLOCYTHEREIS Anderson, 1964

Description. Carapace semicircular to sub-ovate in lateral view with straight ventral margin. Lateral surfaces ornamented by regular pattern of shallow punctae. Slight marginal compression anteriorly, dorsally and posteriorly. Muscle scar pattern consists of a vertical row of four adductors, no other scars observed. Hingement weak, possibly lophodont.

Remarks. This genus is monotypic and its suprageneric placement is uncertain. It is here placed questionably within Limnocytheridae due to its thin shell, muscle scar pattern and weak hingement. However, its occurrence with marginal marine taxa suggests that this genus was capable of withstanding higher salinities than are most modern limnocytherids.

Allocythereis combrookensis Anderson, 1964
Plate 19, figure 7

Description. Monotypic, as for genus.

Remarks. The species probably lived in brackish water or a marginal marine environment.

Distribution. A rare component of collections from the Cotham Member in Combrook railway cutting and Long Itchington Quarry in Warwickshire. Unknown from other Penarth Group horizons. *A. combrookensis* occurs in the upper Rhaetian of eastern Germany.

Family PROTOCYTHERIDAE Ljubimova, 1955
Genus EKTYPHOCYTHERE Bate, 1963

PLATE 19

Description. Anterior and posterior margins compressed. Mid-valve region inflated and ornamented with triangular series of weak to coarse primary ribs with common secondary cross-ribs. Dorsal margin generally straight. Greatest height near anterior margin.

Remarks. *Coronocypris coronula* Anderson, 1964 was erected on the basis of a single small, fragile carapace from the 'Watchet Beds' (=upper Langport Member) at Lavernock Point, South Glamorgan. This carapace appears to be a juvenile *Ektyphocythere* of an undetermined species. *Ektyphocythere* ranges from the Rhaetian to Middle Jurassic.

Ektyphocythere cookiana (Anderson, 1964)
Plate 19, figure 8

Description. Carapace small, adult length 0·45–0·5 mm. Distinctive pattern of concentrically arranged coarse ribs and smooth, compressed marginal areas on lateral surface. Left valve larger than right. Greatest height in front third of valve.

EXPLANATION OF PLATE 19

rt=right; lt=left; car.=carapace; lat.=lateral; RV=right valve; LV=left valve.

Fig. 1. *Ogmoconchella martini* (Anderson, 1964); holotype, GSM Mik (J) 280001; Westbury Formation; Plattlane Borehole, Whixall, Shropshire; car. rt lat.; ×95.

Fig. 2. *Ogmoconchella bristolensis* (Anderson, 1964); holotype, GSM Mik (J) 284001; Langport Member; Lighthorne Quarry, Warwickshire; car. rt lat.; ×87.

Fig. 3. *Cytherella plattensis* Anderson, 1964; holotype, GSM Mik (J) 287001; Langport Member; Lavernock, South Glamorgan; car. lt lat.; ×67.

Fig. 4. *Cytherelloidea* cf. *pulchella* Apostolescu, 1959; GSM Mik (J) 289001; Langport Member; Lavernock, South Glamorgan; car. lt lat.; ×90.

Fig. 5. *Bairdia* sp.; AS collection; top of Cotham Member; Long Itchington Quarry, Warwickshire; RV, fragment; ×58.

Fig. 6. *Lutkevichinella fastigata* Ainsworth, 1989*b*; TCD 28226; Rhaetian; Donegal Basin; car. lt lat.; ×77.

Fig. 7. *Allocythereis combrookensis* Anderson, 1964; AS collection; top of Cotham Member; Long Itchington Quarry, Warwickshire; LV; ×100.

Fig. 8. *Ektyphocythere cookiana* (Anderson, 1964); holotype, GSM Mik (J) 276001; Westbury Formation; Plattlane Borehole, Whixall, Shropshire; car. lt lat.; ×100.

Fig. 9. *Rhombocythere penarthensis* Anderson, 1964; holotype, GSM Mik (J) 272001; Westbury Formation; Seven Sisters Bay, Penarth, South Glamorgan; car. rt lat.; ×57.

Fig. 10. *Darwinula liassica* Jones, 1894; GSM Mik (J) 270001; Penarth Group?, Bristol?; car. rt lat.; ×85.

Remarks. This form is probably the earliest known species of *Ektypho-cythere*. Anderson (1964) erected a new genus, *Klinglerella*, to accommodate this species, but it is considered a synonym of *Ektyphocythere* Bate.

Distribution. Only known from the Westbury Formation of the Plattlane Borehole, Whixall, Shropshire (Anderson 1964) and from Hampstead Farm Quarry, Chipping Sodbury, Gloucestershire (Boomer 1991*a*).

Family *Incertae sedis*
Genus RHOMBOCYTHERE Anderson, 1964

Description. Carapace ovate to sub-quadrate. Anterior and ventral margins broadly rounded, dorsal margin straight to slightly convex, posterior margin square, truncated. Valve margins may possess carinae or weak inflations, particularly dorsally and ventrally. Lateral surfaces smooth to weakly punctate. Adult length 0·55–0·85 mm.

Remarks. The species found associated with *Rhombocythere* suggest that it preferred brackish water conditions. The taxonomic affinities of *Rhombocythere* are uncertain, although its muscle scars and hingement place it within Cytheroidea.

Distribution. The genus is known only from the Upper Triassic of North-West Europe, especially Britain and Germany.

Rhombocythere penarthensis Anderson, 1964
Plate 19, figure 9

Description. Carapace large, rhombic in lateral outline. Distinct dorsal and ventral carinae, the latter terminating in blunt spine posteriorly. Lateral surfaces weakly punctate; punctae broad, circular.

Remarks. One of the few ornate ostracods from the Penarth Group and therefore easily recognizable. However, there is no similarity to Liassic forms and its taxonomic relationships are unclear.

Distribution. Common in the Westbury Formation about 1–2 m above the base at Penarth and Lavernock (Anderson 1964).

Superfamily DARWINULOIDEA Brady and Norman, 1889
Family DARWINULIDAE Brady and Norman, 1889
Genus DARWINULA Brady and Robertson, 1885

Remarks. *Darwinula* has a range from Upper Palaeozoic to Recent and is considered to inhabit freshwater to very low salinity environments.

Darwinula spp.
Plate 19, figure 10

Description. Characterized by an elongate carapace lacking ornament and marginal features. Posterior markedly higher and more inflated laterally than anterior. Length of adult within range 0·6–1·0 mm.

Remarks. *Darwinula* is very largely confined to freshwater habitats but can tolerate slightly saline conditions. Specific differentiation is based on often minor variations in lateral outline and degree of carapace inflation. Due to the difficulty attending the identification of individual species no attempt has been made here to differentiate *Darwinula* below generic level, despite earlier discrimination of two species from the Penarth Group (*D. major* Jones, 1894 and *D. liassica* Jones, 1894) and two from its possible correlative offshore, (*D. rara* Ainsworth, 1989*b* and *D. hettangiana* Ainsworth, 1989*b*). Two further species, *D. globosa* Jones, 1894 and *D. stricta* Jones, 1894 are recorded from the Linksfield erratic in Morayshire (Anderson 1964), but proven correlation with the Penarth Group is lacking for this sequence, and the material is poor.

Distribution. Often abundant in assemblages from higher 'non-marine' levels of the Penarth Group (Anderson 1964).

OTHER OSTRACOD SPECIES

Anderson (1964), in the only comprehensive taxonomic survey of Penarth Group ostracods, erected 21 new species and five new genera and described several other species. Of these, eight species plus *Darwinula* spp. are dealt with herein and a further five species are also discussed. One species not in Anderson (1964) is also described. This leaves several other taxa described by Anderson (1964) whose validity is in doubt, based as they are on few or even single specimens from one locality. Many have no subsequent records, but several were noted in the Cotham Member at localities in Warwickshire by Edmunds *et al.* (1965), close to the site of Anderson's original records. The species are *Eocythereis vertebrale*, *Metacytheropteron? striatulum* and *Klinglerella? cellula*, which are all distinctive ornate forms, and *Pontocypris grandis* and *Pontocypris richardsoni*, which are elongate and smooth. Of the ornate forms *E. vertebrale* would now be reassigned to the genus *Eucytherura*, which has a range of Triassic–Recent. Its occurrence in the Cotham Member makes it the earliest cytherurid recorded from Britain. *Metacytheropteron? striatulum* is now referred to the genus *Timiriasevia* and *K.? cellula* is a juvenile of an unidentified species of *Ektyphocythere*.

Class BRANCHIOPODA Latreille, 1817
Subclass DIPLOSTRACA Gerstaecker, 1866
Order CONCHOSTRACA Sars, 1867
Suborder SPINICAUDATA Linder, 1945

Superfamily CYZICOIDEA Baird, 1849
Family CYZICIDAE Stebbing, 1910
Genus EUESTHERIA Depéret and Mazeran, 1912

Remarks. Jones (1863) initially assigned a species of small bivalved crustacean from the Cotham Member to *Estheria* Rueppell. Raymond's (1946) revision recommended reassignment to *Euestheria* Depéret and Mazeran; all modern records employ this name. But Tasch (1969) subsumed *Euestheria* in *Cyzicus* Audouin, creating a subgenus *Cyzicus* (*Euestheria*). The creation of a subgenus is not supported here and the Penarth Group conchostracans are retained in *Euestheria*.

Euestheria minuta (von Alberti, 1832)
Plate 20, figures 1–4

Description. Carapace variable, small, sub-ovate. Hingeline straight. Umbo placed anteriorly; anterior margin falls away abruptly from umbo. Reticulate ornament of small polygons occurs between numerous, prominent concentric growth rings (Pl. 20, figs 3–4).

Remarks. The authorship of *E. minuta* has, at various times, been attributed to von Dechen (1832), von Alberti (1832, 1834), von Zieten (1833) and Goldfuss (1834). The first formal description was by von Alberti (1834), but he actually published the name earlier, in 1832, attributing the species to Goldfuss. However, Goldfuss's description (1834) was published after that of Alberti. In his German version of De La Beche's *Handbook of geology*, von Dechen (1832) listed the species without description or illustration as *Posidonia minuta* von Alberti. Von Zieten figured it in 1833, citing Alberti as the authority, but gave no description. Thus the authority and validity is insecure, since the first mention (von Alberti 1832) was without description or figure. However, the use of *minuta* and its attribution to Alberti have become so ingrained that the practice is maintained here.

Jones (1863) gave the name *Estheria minuta* var. *brodieana* to the first specimens recorded from the Penarth Group, citing smaller size and 'relatively smaller reticulation' to distinguish them from *E. minuta*. However, there seems little reason to perpetuate this separation, as the actual differences from older Triassic forms of the latter species are minor. Raymond (1946) in a three line summary raised the name *brodieana* to species rank and used the species name *Euestheria brodieana* (Jones), on the basis of 'smaller polygons', but gave no formal description. Again, the establishment of a separate species is considered to be unjustified.

Nearly all modern branchiopod crustaceans are restricted to brackish or freshwater systems and there is little to suggest that *E. minuta* inhabited anything other than these environments. Association with marine forms is rare and is assumed to be the result of post-mortem mixing. Normally, if found with other taxa, the associates are non-marine ostracods or the brackish/freshwater aquatic liverwort *Naiadita*. The most characteristic occurrence of *E. minuta* is as monospecific 'floods', typically in the '*Estheria* Bed' of central and south-central England and in the distinctive micritic nodules of the Cotham Member further north. This compares with the ecology of modern branchiopods, which particularly favour environments such as ephemeral lakes or pools which exclude potential predators or competitors, or indeed any other animals. As the species is restricted to the Cotham Member, it has some value in local correlation and is also an excellent environmental indicator.

Distribution. Restricted to the Cotham Member, but widely recorded from that unit. Common in the south and central Midlands, less so in southern Britain south of Bristol.

Class MALACOSTRACA Latreille, 1806
Subclass HOPLOCARIDA Calman, 1904
Order DECAPODA Latreille, 1803

Remarks. Decapod crustaceans are very rare components of Penarth Group faunas as body fossils, although they may be represented by accompanying ichnofaunas. Material so far described comprises an aegerid shrimp and a species of the somewhat enigmatic, possibly eryonid, genus *Tropifer*. Undescribed remains from the Westbury Formation include cumaceans, isopods and eumalacostracan fragments (Förster and Crane 1984).

Suborder DENDROBRANCHIATA Bate, 1888
Superfamily PENAEOIDEA Rafinesque, 1815
Family AEGERIDAE Burkenroad, 1963
Genus AEGER Münster, 1839
Aeger gracilis Förster and Crane, 1984
Text-figure 8

Description. Carapace about 10 mm long, cylindrical to slightly laterally compressed, smooth, lacking spines. Ventral and posterior margins of carapace accentuated by narrow rim, strongest posteriorly. A number of grooves clearly visible on lateral walls of carapace (Text-fig. 8). Short cervical, almost horizontal hepatic, and shallow antennal grooves arranged in shape of an elongate cross. Post-cervical groove short; branchiocardiac groove longer, sweeping posterodorsally on internal casts. These grooves divide lateral walls of carapace into three spines: the hepatic, lateral and post-antennal spines. Rostrum short, triangular, produced backward as a dorsal keel along anterior dorsal margin. Keel approximately one-third of length of carapace.

PLATE 20

Abdomen more than twice length of carapace, divided into six segments, posteriormost longest. Cuticle punctuated by numerous tiny pits. **Distribution**. Species known from a number of carapaces and abdominal fragments. *A. gracilis* is so far recorded only from the blue and green clays toward the base of the Westbury Formation at Vallis Vale near Frome in Somerset (Copp, *in* Duffin 1982*b*; Förster and Crane 1984).

Infraorder PALINURA Latreille, 1803
Family ERYONIDAE de Haan, 1841?
Genus TROPIFER Gould, 1857
Tropifer laevis Gould, 1857
Text-figure 9A

Description. *Tropifer laevis* is known from the carapace, abdomen and parts of limbs. Carapace approximately rectangular, *c.* 13 mm long. Lateral and posterior margins accentuated by slight rim. Three longitudinal ridges, one median, other two laterally placed, divide carapace into four sections. An irregular 'V'-shaped cervical furrow developed about half way along carapace, dividing it into anterior and posterior halves. The second of six abdominal somites is largest. All have central tubercle expanded into a ridge in somites IV and V. Telson shield-shaped, finely tuberculate, with two bowed longitudinal ridges dividing surface into three subequal parts.

Remarks. *Tropifer laevis* is known from two specimens contained within coprolites from the bone-bed near the base of the Westbury Formation at Aust Cliff (Gould 1857; Duffin 1978*a*). There is insufficient evidence to determine whether *T. laevis* formed part of the diet of the organism producing the coprolites, or whether it was a coprophage, feeding on the material of the coprolite itself.

Distribution. Basal Westbury Formation bone bed, Aust Cliff, Gloucestershire.

Section EUBRACHYURA de St Laurent, 1980
Subsection HETEROTREMATA Guinot, 1977

EXPLANATION OF PLATE 20

Figs 1–4. *Euestheria minuta* (Alberti, 1832); AS BC/C/9; Cotham Member; Bantycock Quarry, near Newark, Nottinghamshire. 1–2, groups of valves; ×5. 3–4, close-ups of reticulate polygonal ornament; ×100, ×200.

Figs 5–9. *Eolepas rhaetica* (Moore, 1861*b*). 5–8 specimens from BMNH Moore collection; near base of Westbury Formation; Vallis Vale, near Frome, Somerset. 5, rostrum; ×10. 6, tergum; ×5. 7, scutum; ×10. 8, carina; ×10. 9, BRLSI M 58; tergum; ×10.

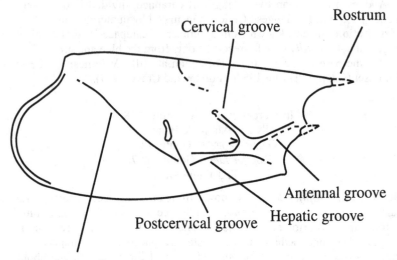

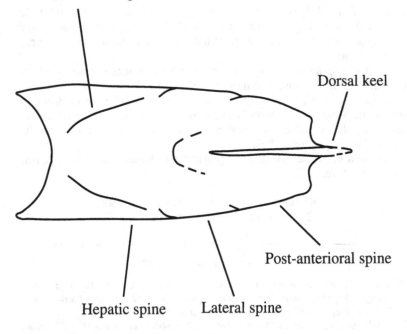

TEXT-FIG. 8. *Aeger gracilis* Förster and Crane, 1984; carapace; lateral and dorsal views; ×8 (after Förster and Crane 1984, p. 460, fig. 4).

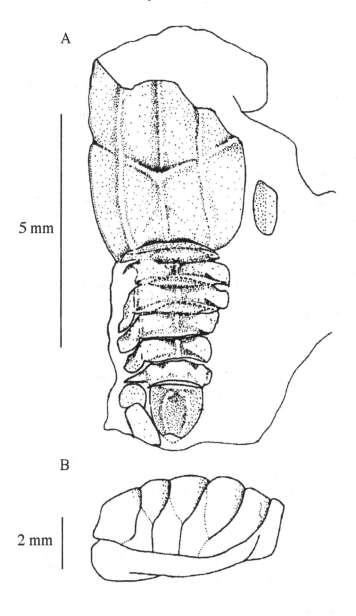

TEXT-FIG. 9. A, *Tropifer laevis* Gould, 1857; lateral view of paratype (after Duffin 1978*a*, p. 180, fig. 2). B, isopod? indet.; lateral view (after Duffin 1978*a*, p. 183, fig. 3A).

Order EDRIOPHTHALMA Leach, 1815
Suborder ISOPODA Latreille, 1817
isopod? indet.
Text-figure 9B

Description. Known from thoracic somites only. Entire structure measures *c.* 5 mm long. Each segment has shallow groove anteriorly which accommodates posterior margin of preceding, overlapping somite. Each somite extends laterally to form pleural extension.

Remarks. This unassigned crustacean was described on the basis of two specimens in a coprolite from Aust Cliff by Duffin (1978*a*). Further specimens are needed to confirm the presence of isopods in the Penarth Group.

Distribution. Basal Westbury Formation bone bed, Aust Cliff, Gloucestershire.

Class CIRRIPEDIA Burmeister, 1834

Remarks. All cirripedes (barnacles) have body and limbs held within an outer integument protected by a number of calcified valves or plates forming the capitulum. The number and disposition of these plates are important taxonomic features. The six main valves consist of paired scuta and terga, plus single carina and rostrum.

Order THORACICA Darwin, 1854
Suborder LEPADOMORPHA Pilsbry, 1916
Family SCALPELLIDAE Pilsbry, 1916
Genus EOLEPAS Withers, 1928
Eolepas rhaetica (Moore, 1861*b*)
Plate 20, figures 5–9; Text-figure 10

Description. Tergae up to 5·6 mm long, elongate, triangular, with pointed apex. Prominent central ridge passes lengthwise down centre of valve, dividing it into two. Growth lines crossed by fine longitudinal striations giving rise to reticulated outer face. Scutum up to 4 mm long, triangular in outline, moderately convex, lacking central ridge of tergum. Carina up to 3 mm long, elongate, narrow, with strongly convex triangular plate with deeply concave inner surface. Rostrum sub-rectangular, shallowly convex, up to 5·5 mm long. Rostrae, scutae and carinae all possess variable amount of reticulate ornament similar to that of tergal plates.

Remarks. The species is most usually represented by fairly small tergal plates. *Eolepas* was attached to the substrate by means of a flexible, muscular peduncle as in extant goose barnacles. Originally allocated to the genus *Pollicepes* by Moore (1861*b*), this is the oldest recorded scalpellid barnacle (Newman *et al.* 1969) and is the type species of *Eolepas*.

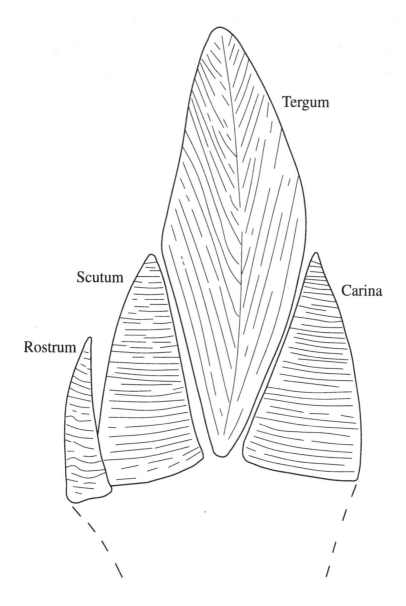

TEXT-FIG. 10. *Eolepas rhaetica* (Moore, 1861*b*); reconstruction.

Distribution. According to Moore (1861*b*, p. 512) *E. rhaetica* 'is not uncommon in the blue clay resting upon the Carboniferous Limestone in the Vallis section' i.e. near the base of the Westbury Formation at a similar horizon to that from which *Aeger gracilis* was recovered. *E.* sp. occurs in the Westbury Formation at Lavernock Point, South Glamorgan (Waters and Lawrence 1987).

10. ARTHROPODS 2: INSECTS

by ED. A. JARZEMBOWSKI

Fossil insects occur sporadically in the various limestone developments in the Penarth Group, especially in west and south-west England, and are occasionally concentrated at certain horizons. The main source of specimens is the distinctive beds of the Cotham Member, including the Cotham Marble. Insects in these beds are often associated with freshwater ostracods, conchostracans and sometimes the aquatic liverwort *Naiadita lanceolata* Buckman, emend. Harris; rarer marine elements may also be present. The environment in which these insects were fossilized probably consisted of marginal freshwater, brackish or estuarine areas where mixing of freshwater, marine and blown-in or transported terrestrial elements could occur. Succeeding Lias Group beds have yielded a richer entomofauna, and at least some of this material from basal levels is of Rhaetian age.

No comprehensive study is available on the Penarth Group insect fauna and older records await confirmation. This in part is due to the fact that Penarth Group insects were considered to be of Early Jurassic age until well into this century, an error that sometimes persists to the present day. The picture is further complicated by the past indiscriminate application of the term 'insect bed' or 'insect limestone' to a number of different beds in the Penarth Group, the Rhaetian portion of the Lias Group and the succeeding Jurassic itself. A definitive species list is therefore beyond the scope of this study, so the possible range of insects has been illustrated via the better documented examples. These examples include taxa from the basal Lias Group (Preplanorbis Beds) as well as the Penarth Group, pending the resolution of stratigraphical problems, and this study more accurately reflects the distribution of insects in the British Rhaetian rather than the Penarth Group *per se*. Some species are known to cross the Triassic/Jurassic boundary as presently conceived. The addition of other groups and new taxa will no doubt follow as research proceeds.

The most distinctive insect fossil remains are wings, especially forewings, although bodies and body parts also occur. Fossils range in size from a few millimetres to several tens of millimetres long. At least nine major groups (orders) of insects are represented, the last five in the text being holometabolous insects i.e. with complete metamorphosis. The remainder are non-holometabolous winged insects.

All figured specimens are held in the collections of The Natural History Museum, London, except the specimen in Text-figure 17c which is held at Bristol City Museum and Art Gallery.

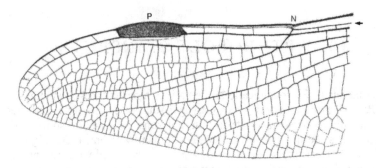

TEXT-FIG. 11. *Diastatommites liassina* (Strickland, 1840); wing; Cotham Member; near Axmouth, Dorset. N=nodus, P=pterostigma, arrow indicates subcostal vein (ScP). Determined by A. Nel, original drawing by R. Coram.

Order ODONATA Fabricius, 1793
(Dragonflies)
Text-figure 11

Remarks. Dragonflies are some of the largest insects found in the British Rhaetian, with both slender archizygopteran (Tillyard 1925, fig. 14) and more robust 'anisozygopteran' species (Text-fig. 11) present. However, no Protodonata are known. Dragonfly wings are distinguished by dense cellular venation, indented anterior margin (at the nodus) and elongate dark cell (pterostigma) beyond the nodus near wing apex. *Diastatommites liassina* (Strickland) of the family Archithemistidae, the 'anisozygopteran' illustrated in Text-figure 11, has a pterostigma well separated from the wing apex and a complete wing length of 70 mm.

Order BLATTODEA Latreille, 1810
(Cockroaches)
Text-figure 12A

Remarks. Fossil cockroaches are uncommon, but the stiffened forewings which are modified as hindwing covers (tegmina) are distinctive. The latter have a dense venation, the veins arising from up to four main stems near the longitudinal axis of the wing (Text-fig. 12A). The stems are curved (weakly sinuous) and the veins may fork towards the wing margin. There is a general lack of crossveins, although intercalary veins may be present; these are distinguished by lacking a basal attachment.

Order DERMAPTERA de Geer, 1773
(Earwigs)
Text-figure 12B

A

B

TEXT-FIG. 12. A, *Rithma liasina* (Giebel, 1856); forewing; BMNH I 12792; Upper Triassic; Strensham, Worcestershire; ×8. B, unnamed dermapteran; elytron; BMNH I 10961; Upper Triassic; Forthampton, Gloucestershire; ×16.

Remarks. Dermaptera are represented by occasional forewing cases which resemble beetle elytra (see below), especially in the possession of several longitudinal ridges (striae). However, they differ from beetle elytra by being obliquely truncate posteriorly with the inner margin longer than the outer margin (Text-fig. 12B). Assignment of specimens exhibiting this morphology to the earwigs is based on the observations of Shcherbakov (pers. comm.).

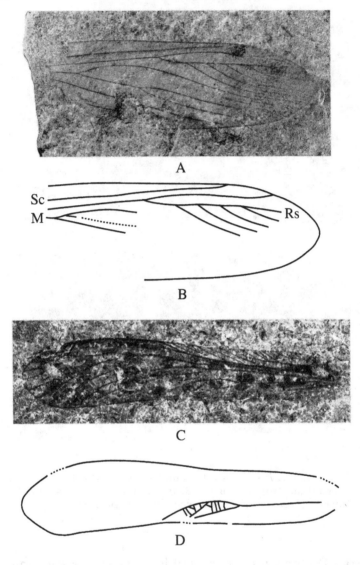

TEXT-FIG. 13. A, *Bintoniella brodiei* Handlirsch, 1939; forewing of female; BMNH I 10554; Upper Triassic; Strensham, Worcestershire; ×3 (after Whalley 1982, p. 147, fig. 5). B, partial vein diagram of forewing of female *B. brodiei* shown in A; ×3. C, *Locustopsis spectabilis* Zeuner, 1942; forewing; BMNH I 49593; Lias Group (lower); Dorset; ×4 (after Whalley 1985, p. 138, fig. 27). D, outline and anterior cubital vein in forewing of *L. spectabilis* shown in C, crossveins included; ×4.

Order ORTHOPTERA Olivier, 1789
('grasshoppers')
Text-figure 13

Remarks. Grasses had not evolved by the Late Triassic, yet both long- and short-horned grasshoppers (suborders Ensifera and Caelifera respectively) are recorded from the British Upper Triassic. The 'horns' are in fact antennae. Short-horned grasshoppers are commonly referred to as bush crickets.

Orthopteran wings are often distinguished by the presence of perpendicular crossveins between the longitudinal veins as in *Bintoniella brodiei* Handlirsch (Whalley 1982, fig. 6; see also Text-fig. 13D). However, these crossveins may be indistinct in fossils (Text-fig. 13A). The well-developed venation and medium to large size of many orthopteran wings (*c*. 20–45 mm long) may cause confusion with dragonflies, but in the former there is no nodus, and the subcostal vein (Sc) runs to the anterior wing margin and not a nodus. In addition, there is never a vein-bound pterostigma, and the radial sector vein (Rs) is pectinately branched posteriorly (Text-fig. 13B). In bintoniellids (extinct Ensifera), the median vein (M) divides distinctly into three branches (Text-fig. 13B).

Locustopseidae (extinct Caelifera) have elongate forewings with concave anterior and posterior margins resembling Elcanidae which are narrower and smaller than Bintoniellidae (Text-fig. 13C). However, they may be distinguished from both by the distinctive tripartite branching of the anterior cubital vein, CuA (Text-fig. 13D). Rhaetian ensiferan wings may show sexual dimorphism which is well-marked in the cricket-like extinct Haglidae and Protogryllidae which sometimes occur (see Zeuner 1939).

Order HEMIPTERA Linnaeus, 1758
(Bugs)
Text-figure 14

Remarks. In the entomological sense, bugs are insects characterized by the presence of piercing and sucking mouthparts in the rostrum, normally used to draw plant fluids and sometimes blood. Both suborders of Hemiptera, i.e. Homoptera and Heteroptera, are present in the British Rhaetian, the names referring to the uniform and divided forewings in the suborders respectively. Heteroptera are sometimes referred to as the true bugs, and the forewings are divided into a tough basal region and a membraneous apical portion. The single heteropteran recorded from the British Rhaetian, '*Pachymerus*' *zucholdi* Giebel (Pentatomorpha: Pachymeridiidae), was revised by Popov *et al.* (1994), as was an extinct coleorrhynchous homopteran, *Archicercopis* (Progonocimicidae).

Text-figure 14 illustrates a cicada-like homopteran from the Cotham

TEXT-FIG. 14. *Liassocicada ignotata* (Brodie, 1845); female; BMNH I 11244; Upper Triassic; Strensham, Worcestershire; x 6. Arrow indicates stylets of rostrum which is 14 mm long (after Whalley 1983, p. 144, fig. 8).

Member with characteristic long mouthparts for probing plants. The rostrum is slung beneath the body, extending from the head (right) to the genitalia (left).

Order NEUROPTERA Linnaeus, 1758
(Lacewings)
Text-figure 15

Remarks. As the name implies, wings of lacewings are membranous, often with a dense venation. Usually three longitudinal veins run more or less parallel to each other in the fore- and hindwings, a short distance behind the anterior margin (Sc, R1 and Rs in Text-fig. 15B, the 'vena triplica'). Branches of the main veins often end in smaller dichotomous branches near the wing margin ('end-twigging', see Text-fig. 15C).

Neuropteran wings from the Penarth Group may be large or small. In the large *Megapolystoechus magnificus* Tillyard (Text-fig. 15A), which possesses a wing length of more than 40 mm, there are numerous (>24) pectinate branches of Rs joined by crossveins basally; the vena triplica converge but do not meet near the wing apex. In the small *Archeosmylus complexus* Whalley (Text-fig. 15C), which has a wing length of c. 10 mm, there are only seven to ten branches of Rs not joined by crossveins; Sc and R1 are fused apically in the vena triplica.

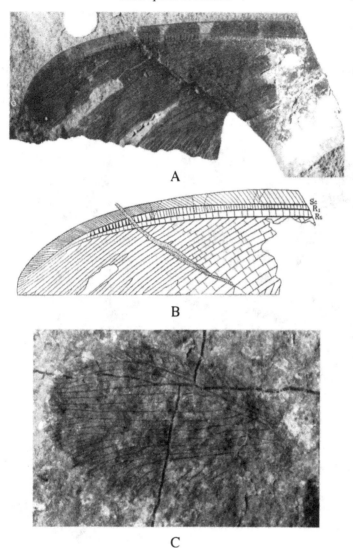

A

B

C

TEXT-FIG. 15. A, *Megapolystoechus magnificus* Tillyard, 1933 (Mesopoly-stoechotidae); forewing; BMNH I 11027; Upper Triassic; Strensham, Worcestershire; ×3 (after Whalley 1988, p. 53, fig. 5a). B, venation diagram of anterior half of forewing of *M. magnificus* shown in A; x 3 (after Tillyard 1933, fig. 1). C, *Archeosmylus complexus* Whalley, 1988 (Permithonidae); wing; BMNH I 11412; Lias Group (upper); near Dumbleton, Gloucestershire; ×8 (after Whalley 1988, p. 51, fig. 1).

Order COLEOPTERA Linnaeus, 1758
(Beetles)
Text-figure 16

Remarks. Beetle forewings are modified into distinctive seed-like wing cases or elytra. These are usually elongate, rounded in cross section, truncate near the base and asymmetrically pointed at the opposite end.

TEXT-FIG. 16. A, *Holcoptera schlotheimi* (Giebel, 1856); paired elytra in rest position; BMNH I 53990; Lias Group (lower); Dorset; x 12·5 (after Whalley 1985, p. 175, fig. 82a). B, *Holcoptera giebeli* (Handlirsch, 1906); left elytron; BMNH I 53989; Lias Group (lower); Dorset; ×10 (after Whalley 1985, p. 177, fig. 85a).

Elytra may be smooth or ornate, the latter often with stipples (punctae), e.g. *Metacupes harrisi* Gardiner. The most distinctive elytra of Penarth Group beetles, as in the Lias Group, are those of the genus *Holcoptera*, which has striped light and dark markings. The elytron of *Holcoptera schlotheimi* (Giebel) is 5–6 mm long with four dark stripes (Text-fig. 16A). The elytra of *H. giebeli* (Handlirsch) are 11–14 mm long with five dark stripes (Text-fig. 16B).

Beetles are common among Penarth Group insect fossils and there are probably many taxa awaiting description. The family classification of Mesozoic beetles is problematical, e.g. genus *Holcoptera* is family uncertain and *Metacupes* is a questionable cupedid. Local records of the collective groups *Buprestites* and *Pterostichites* from the Upper Triassic of Gloucestershire and Leicestershire are believed to exist, and other fossil and even Recent genera have been claimed, but such reports need extensive verification and evaluation of museum collections. The discovery of intact bodies may provide useful additional systematic characters.

Order MECOPTERA Packard, 1886
(Scorpionflies)
Text-figure 17A–B

Remarks. The common name is derived from the resemblance of the genitalia of Recent British males to a scorpion's sting. Mecoptera are represented by several species in the British Upper Triassic, of which the genus *Orthophlebia* (extinct family Orthophlebiidae) is the most common and usually represented by the type species *O. liassica* Mantell. Uncommon scorpionflies include *Protorthophlebia* (Orthophlebiidae) and *Protobittacus* (Neorthophlebiidae) (see Tillyard 1933).

Wings of Orthophlebiidae are 12–18 mm long with a long subcostal vein (Sc), a radial vein (R) diverging from Sc with six to eight branches and (usually) five medial veins (M) (Text-fig. 17B). In *Orthophlebia liassica* the wings are 12–17 mm long with six or seven branches of Rs (R minus R1) (Text-fig. 17B), whereas in *O. gigantea* from the Upper Triassic of Worcestershire there are eight branches of Rs and wing length exceeds 18 mm.

Order DIPTERA Linnaeus, 1758
(True flies)
Text-figure 17C

Remarks. True flies are uncommon in the Penarth Group and also in the succeeding Lias Group. An unpublished eoptychopterid is known from Gloucestershire and a solitary, unnamed species is illustrated in Text-figure 17C.

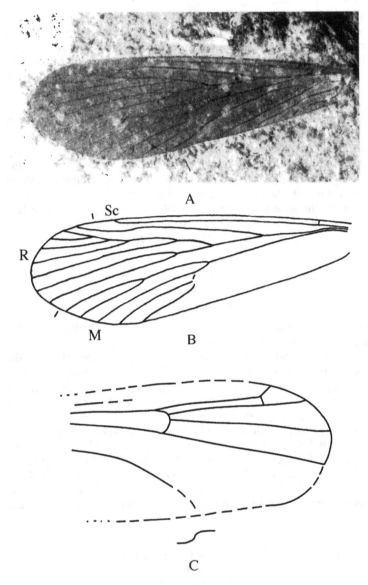

TEXT-FIG. 17. A, *Orthophlebia liassica* (Mantell, 1844); forewing; BMNH I 10761; Upper Triassic; Gloucestershire; ×6. B, partial venation of forewing of *O. liassica* shown in A; ×6. C, unnamed dipteran; forewing; Bristol City Museum and Art Gallery, Jarzembowski collection Cd 4127a-b; Cotham Member; Aust Cliff, Gloucestershire; ×60. Original drawing by P. Stevenson.

A

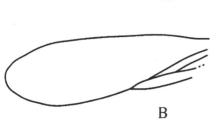

B

TEXT-FIG. 18. A, *Necrotaulius furcatus* (Giebel, 1856); adult; BMNH I 11522; Upper Triassic; Gloucestershire. B, outline of forewing of *N. furcatus* illustrated in A, showing looped anal veins. Both ×8.

Two-winged or true flies possess only forewings, the hindwings being highly modified as balance organs (halteres). The wings are small–medium sized and membranous with reduced venation and can be distinguished by the curved cubital vein having a step-like profile in cross-section (Text-fig. 17C) because it consists of a strong anterior cubitus with raised relief and a faint but parallel posterior cubitus with depressed relief in top view.

Order TRICHOPTERA Kirby, 1815
(Caddisflies)
Text-figure 18

Remarks. The winged adults are normally associated with fresh water in which the larvae live. British Rhaetian caddisflies are small insects belonging to the extinct family Necrotauliidae (Text-fig. 18A), members of which may have been tolerant of brackish water conditions. Trichoptera have a reduced wing venation with distinctive looped anal veins near the inner edge of the posterior margin in the forewing (Text-fig. 18B).

11. ECHINODERMS

by ANDREW SWIFT

There have been very few systematic studies of echinoderms from the Penarth Group. Wright (1880) and Hess (1965) described the ophiuroid *Aplocoma damesii* (Wright), which is common in the Westbury Formation but not recorded from higher parts of the succession. On the other hand, echinoid fossils of any degree of completeness or size are very much rarer. Wright (1861, p. 457) described the regular echinoid *Hemipedina tomesii* and figured a specimen from the 'White Lias' (=Langport Member) of Stoneythorpe, Warwickshire. This specimen remains the best and indeed probably the only moderately complete example recovered from the Penarth Group (Pl. 21, fig. 1). Smith (1990) presented a detailed review of all known species of echinoids recovered from the Lower Triassic to the Hettangian Stage of the Lower Jurassic and recorded only two species in the Penarth Group, *Diademopsis herberti* (Agassiz and Desor) (=*Hemipedina tomesi* Wright) and *D. serialis* (Agassiz). Smith (1990, pl. 3, fig. 2) refigured Wright's (1861) specimen of *D. herberti* from Stoneythorpe. *D. serialis* is only represented in the Penarth Group by fragmentary remains. Other published information concerning echinoids in the Penarth Group is mostly in the form of records of indeterminate spine and test fragments (e.g. Warrington and Ivimey-Cook 1990), but the nature of such material leaves little scope for taxonomic assessment.

Such a dearth of specific information suggests that echinoderms were minor components in the Penarth Group biota, but evidence from micropalaeontological analysis indicates otherwise. Disarticulated spines, plates and other test fragments of echinoids (Pl. 21, figs 2–4) are common in residues which result from the disaggregation of Penarth Group rocks, together with ophiuroid remains (Pl. 21, figs 12–15), holothurian sclerites (Pl. 21, figs 16, 19–21) and other indeterminate echinoderm remains (Pl. 21, figs 17–18). These occurrences and the common occurrence of ophiuroids and echinoid and holothurian debris in the Westbury Formation, indicate that echinoderms formed a significant portion of the Penarth Group biota throughout its marine phases of deposition.

Recent echinoderms inhabit a wide range of environments in modern oceans, but are usually confined to waters with normal marine salinities and have little tolerance of salinity variations. However, within this limit, they have colonized every depth zone in the oceans from the intertidal zone to abyssal plains and trenches. Echinoids and asteroids are typical of shallow waters, but holothurians and ophiuroids, as well as inhabiting shallow water, have modern species adapted for life in the deepest marine settings, and are commonly the dominant large invertebrates of these environments. If ancient echinoderms were similar in salinity tolerance to their modern representatives, then it is probable that for significant portions of time during deposition of the Penarth Group, approximately stenohaline conditions prevailed.

Another major group of echinoderms, the crinoids, are apparently absent from the Penarth Group. Moore's (1861*b*, p. 512) record of 'joints of a small Encrinite' in the Westbury Formation clays at Vallis Vale (Frome) may refer to derived material. Moore (1861*b*) gave no description or details of his finds, and the disposition of the basal Westbury beds at Vallis Vale, which are associated with conglomerates rich in derived clasts of Carboniferous Limestone would suggest that the specimens were reworked from Carboniferous beds. Teeth and scales of Carboniferous fishes have been recorded from similar Penarth Group horizons (see Chapter 14). Crinoids, whose low diversity modern representatives inhabit a wide range of environments in marine waters, are amongst the commonest of fossils in the succeeding Jurassic and have a long fossil record dating back to the Middle Cambrian. Since salinity, temperature and available light would at times during deposition of the Penarth Group have been within the tolerances of crinoids, the lack of indigenous examples is surprising. However, crinoids in the modern seas are not common in oxygen-deficient settings or in very shallow water, so they may have been unable to colonize the Westbury Formation environments for the former reason, and the Lilstock Formation waters for the latter.

The specimens figured on Plate 21 are held in The Natural History Museum, London (fig. 1); in the A. Swift collection, Department of Geology, University of Leicester (figs 2–4, 10–21) and in the New Walk Museum, Leicester (figs 5–9).

SPECIES DESCRIPTIONS

Phylum ECHINODERMATA Bruguière, 1791–92
Class ECHINOIDEA Leske, 1778
Subclass EUECHINOIDEA Bronn, 1860
Infraclass ACROECHINOIDEA Smith, 1981
Cohort DIADEMATACEA Duncan, 1889
Order PEDINOIDA Mortensen, 1939
Family PEDINIDAE Pomel, 1883
Genus DIADEMOPSIS Desor, 1855
Diademopsis herberti (Agassiz and Desor, 1847)
Plate 21, figure 1

Description. Test small–medium, low hemispherical or rotular. Ambulacral and interambulacral plates narrow, primary tubercle central, scrobicular ring of tubercles extends to plate boundaries. Ambulacral pores uniserial. Ambulacral plates near mouth have three pore pairs; those farther away simple but with primary tubercule every third plate. Inter-ambulacral plates low, broad, bearing more than one series of enlarged tubercles which are similar in primary and secondary series. (Based on Fell (*in* Moore 1966) and Smith 1990).

Remarks. Test fragments and spines which are probably attributable to *D. herberti* (Pl. 21, figs 2–4) are common in both the Westbury Formation and the Langport Member, especially the latter, but as far as is known the specimen illustrated here and on which systematic knowledge of the species in Britain is based, is the only fossil with any degree of complete-ness. As *Hemipedina tomesii*, the species has been recorded on a few occasions, e.g. at Print Hill in Warwickshire by Woodward (1893). Richardson (1912) recorded the same species from the same locality, claiming a Langport Member source for his specimen(s?). Neither Woodward (1893) nor Richardson (1912) gave any description or illus-tration and the nature and preservation of the specimens cannot be verified. Old *et al.* (1987) and Warrington *et al.* (1994) referred material from the Langport Member of Warwickshire and the Lilstock Formation of Somerset to *Diademopsis*, but made no specific assignment.

The Diadematacea make their first undoubted appearance in the Upper Triassic and become common in the Jurassic.

Distribution. Widespread and common in the form of test fragments and spines, especially in the Langport Member.

Diademopsis serialis (Agassiz, *in* Leymerie, 1838)

Remarks. Only rare test fragments of this species have been noted and they are found in association with the remains of *D. herberti* (Smith 1990). The two species may be distinguished by differences in the tuberculation e.g. the tubercles of the interambulcral plates of *D. serialis* are of similar size in both the primary and secondary series, whilst the secondary series of *D. herberti* are smaller than the primaries. For more detail on the differences and the distribution of *D. serialis*, see Smith (1990).

Class OPHIUROIDEA Gray, 1840*a*
Order OPHIURIDA Müller and Troschel, 1840
Suborder CHILOPHIURINA Matsumoto, 1915
Family OPHIURIDAE Lyman, 1865
Genus APLOCOMA d'Orbigny, 1852
Aplocoma damesii (Wright, 1874)
Plate 21, figures 5–11

Description. Small, disc up to 7 mm in diameter, arms up to 25 mm long. Disc upper surface convex, raised where arms join, with concentric plates.

PLATE 21

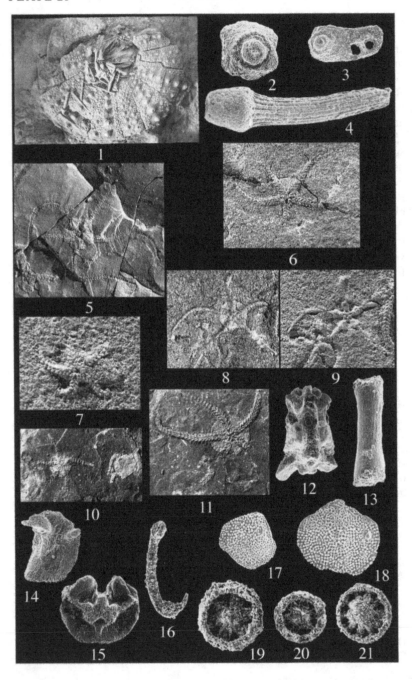

Arms long, thin, tapering from disc to fine point. Upper surface of arms ornamented with interlocking plates, diminishing in size distally from disc. Lower surface of arms has double row of articulating segments.

Remarks. All brittle-stars identified until recently from the Westbury Formation were assigned to *Ophiolepis damesii* Wright, which was the name given to the first British specimens (Wright, *in* Harrison 1876). Wright (1874) based his first description of *O. damesii* on specimens from Hildesheim (Germany). He subsequently recognized comparable specimens from Garden Cliff, Gloucestershire and Spinney Hills, Leicestershire, and provided a very detailed description of *O. damesii* (Wright 1880). Hess (1965) published an overview, with systematic revisions of Triassic ophiuroids, and concluded that, at most, only four species were valid, all belonging to *Aplocoma*. Hess (1965) examined specimens from Aust Cliff and Spinney Hills but was not convinced of their relationship to his accepted species and instead assigned them to *Aplocoma* sp., a generic assignment followed here. Hess (1965) regarded

---

EXPLANATION OF PLATE 21

cop.=copy of

Fig. 1. *Diademopsis herberti* (Agassiz and Desor, 1847); BMNH E 8664; Langport Member; Stoneythorpe, Warwickshire; cop. Smith 1990, pl. 3. fig. 2; ×1·5.

Figs 2–4. Echinoid fragments; top of Cotham Member; Long Itchington Quarry, Warwickshire. 2, tubercle; AS 21/15; ×35. 3, ambulacral plate showing tubercle and paired pores; AS 29/10; ×35. 4, spine; AS 21/16; ×20.

Figs 5–11. *Aplocoma damesii* (Wright, 1874); Westbury Formation. 5–9, Spinney Hills, Leicester. 5, NW 37'1921b; oral view; ×1. 6, specimen from slab NW 37'1921c; aboral view; ×2·5. 7, another specimen from slab NW 37'1921c; aboral view; ×4. 8 (direct lighting) –9 (oblique lighting), another specimen from slab NW 37'1921c; oral view; AS 12/24; ×2·5. 10–11, Bantycock Quarry, Newark, Nottingham. 10, oral view; ×2·5. 11, AS 12/30; oral view; ×2·5.

Figs 12–15. Ophiuroid fragments. 12–13, top of Cotham Member; Long Itchington Quarry, Warwickshire; lateral view. 12, vertebra; AS 23/19. 13, fragment; AS 29/9; lateral view. 14, ambulacral plate; AS 28/14; Westbury Formation; Bantycock Quarry, Newark, Nottingham. 15, vertebra; AS 32/38; top of Westbury Formation; Lavernock Point, South Glamorgan; end view. All ×35.

Fig. 16. *Achistrum* sp.; AS 31/32; Langport Member; Long Itchington Quarry, near Southam, Warwickshire; ×50.

Figs 17–18. Perforated 'plates'; AS 22/18, AS 22/19; Westbury Formation; Bantycock Quarry, Newark, Nottingham; ×50.

Figs 19–21. *Theelia* spp.; Langport Member. 19, AS 31/30; six-spoked form; Long Itchington Quarry, near Southam, Warwickshire. 20, AS 32/33; six-spoked form; Pinhay Bay, Devon. 21, AS 31/31; ten-spoked form; Long Itchington Quarry, near Southam, Warwickshire. All ×100.

O. damesii as a *nomen dubium*. However, Wright (1874) apparently erected the species correctly and later added much further detail complete with illustrations (Wright 1880); thus it seems reasonable to continue use of *damesii* as the specific name. Further taxonomic investigation may reveal that corrections are necessary, but that is more appropriately undertaken elsewhere.

Wright (1880, p. 457) produced meticulous drawings of two of his British specimens. Horwood (1916) also figured good specimens from the Westbury Formation of Glen Parva, Leicestershire (Horwood 1916, pl. 17, figs 3–4; herein Pl. 21, figs 5, 8–9, together with two other specimens from the slab containing Horwood's fig. 4), but mistakenly attributed them to the Lower Lias in his plate caption. However, detailed comparison of fossil material with published descriptions can be difficult as some British specimens are very small, fragmentary and poorly preserved in the form of structureless calcitic replacements, or compressions. Horwood (1916) mentioned both large and small forms of ophiuroid and suggested that they might be dimorphic, whereas Harrison (1876, 1880) believed that there were two species. Two specimens collected recently from Bantycock Quarry, Nottinghamshire are figured (Pl. 21, figs 10–11) and are representative of what the collector can expect to find.

As well as the articulated specimens of *A. damesii* described above, residues prepared for microfossil analysis frequently contain disarticulated ophiuroid material, examples of which are illustrated (Pl. 21, figs 12–15). It seems reasonable to assume that all or most of this material would be the remains of specimens of *A. damesii*. However, it is possible that some elements may be derived from asteroids ('starfish').

In addition to whole body ophiuroid fossils, trace fossils recording ophiuroid burrowing traces are illustrated here for the first time (Chapter 16, Pl. 31, fig. 2; Text-fig. 31L).

Distribution. Known only as macrofossils from the Westbury Formation. Of widespread occurrence, but confined to particular horizons, presumably those where optimum marine conditions prevailed. Associations of several individuals, such as that found by Horwood (1916) at Spinney Hills, Leicester, may reflect mass mortality events triggered by rapid onset of inimicable conditions.

Class HOLOTHUROIDEA de Blainville, 1834

Remarks. Holothurians are represented in the fossil record almost exclusively by their sclerites, which are usually the only mineralized elements capable of preservation. These, in part, aid in supporting and stiffening the otherwise soft body of 'sea-cucumbers' and assume a wide variety of shapes, which have proved difficult to describe scientifically. Many of these morphotypes are reminiscent of familiar objects, and thus their gross morphologies have acquired a set of colloquial descriptive terms, such as 'tables', 'anchors', 'hooks', 'wheels', etc. Assigning these forms is difficult as many taxa share similar characteristics and only two long ranging form genera can be discriminated with any certainty from the Penarth Group: *Achistrum* and *Theelia*. Perforated 'plates', 'tables' and calcareous 'rings' also occur rarely, but classification is difficult. Two perforated 'plates' are illustrated on Plate 21, figures 17–18; these may be of holothurian origin, but doubts have been expressed by many authors (e.g. Gilliland 1992).

Order APODIDA Brandt, 1835
Family ACHISTRIDAE Frizzel and Exline, 1956
Genus ACHISTRUM Etheridge, 1881 emend. Frizzel and Exline, 1956
Achistrum sp.
Plate 21, figure 16

Description. Simple 'hooks' consisting of eye, shank and recurved spear. Crossbars can sometimes be discriminated within the eye.
Remarks. Despite their calcareous composition, many 'hooks' survive acetic acid digestion of limestones. Better preserved examples are recovered from disaggregated mudrocks.
Distribution. Common in the Lilstock Formation, especially the Langport Member, but not yet recovered from the Westbury Formation.

Family SYNAPTIDAE Burmeister, 1837?
Genus THEELIA Schlumberger, 1890
Theelia spp.
Plate 21, figures 19–21

Description. 'Wheels', consisting of rim, spokes and hub.
Remarks. Many species of *Theelia* have been described, distinguished by variations in all three major morphological features. However, it is usually variations in the number and thickness of the spokes which are the main discriminating characteristic. Penarth Group forms are dominantly of the six-spoked type (Pl. 21, figs 19–20) but ten-spoked forms are not uncommon (Pl. 21, fig. 21).
Distribution. As for *Achistrum*.

12. OTHER INVERTEBRATES

by ANDREW SWIFT

CORALS

The corals of the Penarth Group hold much of palaeontological and stratigraphical interest but a systematic revision is long overdue. Much, if not all, of the systematics which remain in place today were established in the late nineteenth and early twentieth centuries. In addition, the most active 'Rhaetic' and Jurassic coral systematist of the day, R. F. Tomes, was a controversial figure who was at odds with many other prominent 'Rhaetic' and coral workers such as Charles Moore, H. B. Woodward and P. M. Duncan. Tomes' belief that the corals of the 'Sutton and Southerndown Beds' of South Wales indicated a 'Rhaetic' age for those beds was particularly controversial. In the light of this, a conservative approach has been adopted here and only reliable data and relatively secure taxa have been included.

Stratigraphically and in evolutionary terms, Penarth Group corals occur at a point early in the radiation of the scleractinian corals, which had replaced all the earlier Palaeozoic orders and were to dominate coral faunas thereafter. The Penarth Group taxa are small, simple in form and generally are scarce. Occurrences in the Westbury Formation can be counted in single figures and it is only in the warm water shallow lagoonal facies of the Langport Member that the corals assume some significance in the fauna, and then only locally.

Only four coral taxa can be identified with confidence from the Penarth Group: *Montlivaltia rhaetica* is by far the most common, *Oppelismilia* sp. and *Thecosmilia* sp. are rare and *Heterastraea rhaetica* is very rare. The true distribution of the colonial *Thecosmilia* sp. may be obscured by it being confused with the solitary *Montlivaltia rhaetica* in cases where only a single corallite is seen, as there is a similarity between the calices of the two species in transverse section. *Oppelismilia* sp. and *Montlivaltia rhaetica* have also been confused with each other and wrongly attributed in the past.

The specimens figured on Plate 22 are held at The Natural History Museum, London (figs 1, 4–13, 15–18), National Museum of Wales, Cardiff (figs 2–3) and the British Geological Survey, Keyworth (fig. 14).

SPECIES DESCRIPTIONS

Phylum COELENTERATA Hatschek, 1888
Class ANTHOZOA Ehrenberg, 1834

Subclass ZOANTHARIA de Blainville, 1830
Order SCLERACTINIA Bourne, 1900
Family STYLOPHYLLIDAE Volz, 1896
Genus HETERASTRAEA Tomes, 1888
Heterastraea rhaetica Tomes, 1903
Plate 22, figures 1–3

Description. Colonial cerioid coral. Corallites irregular, polygonal-walled, in closely packed contact, small to very small, forming small colonies. Septa rather irregular, thickening towards fossula. Calices moderately concave. Septal margins sparingly denticulate.

Remarks. With the exception of the holotype (Pl. 22, fig. 1) from the Westbury Formation in Gloucestershire, the only other known specimens of *Heterastraea rhaetica* from South Wales are notable for their very small size; consequently the species may have been overlooked and may possibly be more common than suspected. The best known specimen remains the holotype, first described by Tomes (1903). This was presented to him by Linsdall Richardson and originated from a temporary excavation into the Westbury Formation at Deerhurst, Gloucestershire. It was obtained from a hard, grey, calcareous sandstone, which contrasted sharply in facies with the surrounding black shale typical of the Westbury Formation. This is probably significant and indicates that, like most hermatypic corals, *H. rhaetica* could only colonize shallow waters within the euphotic zone. In the Late Triassic it is assumed that most if not all corals were hermatypic i.e. that the requirements of their essential symbiotic algae restricted their occurrences to the euphotic zone at depths of less than *c*. 50 m. In fact, it is likely that *H. rhaetica* grew in considerably shallower water within the zone of wave action, hence its occurrence in a calcareous sandstone.

Further evidence of restriction to shallow, well-aerated, relatively high-energy environments is provided by the occurrence of *H. rhaetica* in Late Triassic marginal deposits around Cowbridge in South Wales, which are coeval with parts of the Penarth Group. In this very near shore environment adjacent to a major 'island', small colonies of *H. rhaetica* became established and their remains occur today in calcareous sandstones and limestones (Pl. 22, figs 2–3).

Distribution. Very rare. Westbury Formation at Deerhurst, Gloucestershire and Late Triassic marginal deposits (equivalent to some part of the Penarth Group) around Cowbridge, South Wales.

Genus OPPELISMILIA Duncan, 1868
Oppelismilia sp.
Plate 22, figures 4–7

Description. Fan-shape to trochoid (i.e. basal angle of around 40°), somewhat compressed. Corallites free laterally. Solitary or forms small colonies by extratentacular intracalicular budding. Irregular beaded trabeculate border to wall formed by extended septal spines, which often fuse to produce septal laminae. Columnella absent.

Remarks. *Oppelismilia* sp. is very rare in the Penarth Group and has only been recovered from the two best-known Langport Member exposures on the south coast, Culverhole Point and Pinhay Bay. Only a single occurrence is noted in the literature, a specimen collected on a Geologists' Association field trip in 1955 (Smith 1957). This specimen is held in the collections of The Natural History Museum, London and shows clearly the characteristic thickening by fusion of the septal spines (Pl. 22, fig. 4). Another specimen from the same locality (Pl. 22, fig. 5), shows well the beaded outer margin of the wall. Examples of *Oppelismilia* sp. from the Penarth Group have not previously been illustrated.

Distribution. Only known from the Langport Member at Culverhole Point and Pinhay Bay, Devon.

Family MONTLIVALTIIDAE Dietrich, 1926
Genus MONTLIVALTIA Lamouroux, 1821
Montlivaltia rhaetica Tomes, 1878
Plate 22, figures 8–15

Description. Small, solitary, discoid to horn-shaped. Wall thin, with faint ornament. Calice moderately convex, columella very weak. Septa numerous, thin, extending to outer margin of wall.

Remarks. Some specific distinguishing characters may be masked by the generally poor preservation of the Penarth Group forms, as noted by Duncan (1868), but it is probable that this is the only species of *Montlivaltia* present in the Penarth Group. Very little variability can be discerned in the specimens of *Montlivaltia* in British collections. At times, other species of *Montlivaltia* have been suggested and Duncan (1868), in the first systematic reference to corals from the 'Zone of *Avicula contorta* and the White Lias', stated that there was one species from the 'Zone of *Avicula contorta*' (this information was probably based on the single example from the 'Flinty-bed' at Beer Crowcombe recorded by Moore 1861*b*) and implied that there were three or more from the 'White Lias', but his illustrated 'morphotypes' from Warwickshire probably reflect different states of preservation. One from Punt (=Print) Hill (Duncan 1868) was compared to the Liassic species *M. haimei* Chapuis and Dewalque (this specimen was nominated as the holotype of *M. rhaetica* by Tomes in 1878 and is figured here on Pl. 22, figs 8–9). Richardson (1912) mentioned a species *M. tomesi*, but this is possibly an error and certainly a *nomen nudem*.

PLATE 22

In places *M. rhaetica* is quite common in the Langport Member, especially in quarry exposures in Warwickshire e.g. Long Itchington, but Moore's (1861*b*) specimen of an indeterminate species of *Montlivaltia*? from the 'Flinty-bed' in Somerset is the only record from the Westbury Formation.

Distribution. Confined to the Langport Member, with the exception of a single obscure and incomplete specimen of *Montlivaltia*? sp. from the 'Flinty-bed', low in the Westbury Formation (Moore 1861*b*). Locally common in the thicker Langport Member developments of south Devon and Warwickshire.

Genus THECOSMILIA Milne-Edwards and Haime, 1848
Thecosmilia sp.
Plate 22, figures 16–18

Remarks. *Thecosmilia* is a common Mesozoic genus and its species occur in fasciculate, phaceloid colonies (where the corallites are sub-cylindrical, separate and subparallel). Doubt attends the identification of some of the handful of specimens recovered from the Penarth Group and some may be referable to other genera, including *Montlivaltia* and *Oppelismilia*. Some

---

Figs 1–3. *Heterastraea rhaetica* Tomes, 1903; 1, holotype, BMNH R 17405; Westbury Formation; near 'The Folly', Deerhurst, near Tewkesbury, Gloucestershire; ×2. 2, NMW 67·121·G2; 'base of Upper Rhaetic'; south end of Cowbridge railway cutting, Mid Glamorgan; ×3. 3, NMW 67·121·G80; 'base of Bed 2', 'Rhaetic'; south end of Cowbridge railway cutting, Mid Glamorgan; ×3.

Figs 4–7. *Oppelismilia* sp.; Langport Member. 4, BMNH R 40970; '2·44 m below Gault uncomformity', Culverhole Point, Devon. 5, BMNH R 39290; Culverhole Point, Devon. 6, BMNH R 36920; Pinhay Bay, Devon. 7, BMNH R 45660; lower part of 'White Lias'; Pinhay Bay, Devon. All ×1.

Figs 8–15. *Montlivaltia rhaetica* Tomes, 1878; Langport Member. 8 (original) –9 (cast from original), holotype, BMNH R 13097; Print Hill, Warwickshire. 10, paratype, BMNH R 13109; Harbury, Warwickshire. 11, paratype, BMNH R 13126; Bascote, south-east of Leamington, Warwickshire. 12, BMNH R 13105; group of four calices, Pinhay Bay, Devon. 13, BMNH R 11876; Stockton, south-west of Rugby, Warwickshire. 14, BGS HIC 875; Culverhole Point, Devon. 15, BMNH unnumbered specimen; Long Itchington, Warwickshire. All ×1.

Figs 16–18. *Thecosmilia* sp.; Langport Member. 16, BMNH R 11898; Steven's Hill, Long Sutton, east of Langport, Somerset. 17, BMNH R 11935; Watchet, Somerset. 18, BMNH R 11878; Meer Hill, Loxley, south-east of Stratford-on-Avon. All ×1.

specimens originally assigned to *Thecosmilia* in the collections of The Natural History Museum, London, have subsequently been re-assigned to *Stylophyllopsis* Frech, thought by some authors (e.g. Wells, *in* Moore 1956) to be synonymous with *Oppelsmilia*. Whilst a few specimens (see Pl. 22, figs 16–18) may be true *Thecosmilia*, isolated corallites referred to species of this genus seem more likely to be allied to solitary genera (see particularly Tomes 1884 for several species referred to *Thecosmilia*). Records are few and most data are contained in literature well over a hundred years old (Moore 1861*b*; Duncan 1868; Tomes 1884).

Moore (1861*b*) recorded a bed in the 'White Lias' in a section at Steven's Hill, Long Sutton, Somerset, which contained a large number of corals of a branching form, which Tomes thought were probably referable to a species of the genus *Cladophyllia*. Duncan (1868) intimated the presence of two species of *Thecosmilia* in the Penarth Group and figured both, suggesting that a specimen from Sparkfield was similar to *Thecosmilia terquemi* Duncan, and that the other from Watchet was a 'stunted conico-cylindrical' form. However, both specimens were isolated corallites and no details of the colonial configuration were present. Tomes (1884) identified four species: three from the 'White Lias' of Warwickshire; namely *rugosa*, *hörnesii*, and *confluens* and repeated the record (Moore 1861*b*, p. 492) of the unspecified branching species from Long Sutton in Somerset, which he assigned to *Thecosmilia* sp. (a specimen from Long Sutton is figured here on Pl. 22, fig. 16). The three named species were illustrated (Tomes 1884, pl. 19, figs 1, 7–8) and particularly in the case of *T. rugosa*, the illustration, which shows a colonial coral, suggests assignment to *Thecosmilia* is appropriate (this specimen is re-figured here on Pl. 22, fig. 18).

Subsequent records are few and in the only one of significance Richardson (1911, p. 41) described a massive limestone near the middle of the 'Langport Beds' at Charlton Mackrell containing abundant '*Thecosmilia*(?) *michelini* Terquem and Piette'. Richardson (1911) thought that this bed correlated with Moore's (1861*b*) coralliferous bed at nearby Long Sutton. Arkell (1933, p. 100) repeated Richardson's (1911) record and stated 'near the middle of the series (i.e. Langport Member) at Charlton Mackrell is a coral limestone, 16 in. thick, full of *Thecosmilia*? *michelini* Terquem and Piette, associated with numerous *Ostrea liassica*'. **Distribution**. Only recorded from the Langport Member in Warwickshire and Somerset.

BRACHIOPODS

In contrast with the bivalves, which are relatively diverse and numerous, brachiopods were clearly unsuited to the prevailing conditions during much of the interval when the Penarth Group was deposited. This is demonstrated by the fact that only two lingulate taxa have been recovered, *Lingula* sp., which is very rare, and *Discinisca townshendi*, which is scarce. Both are restricted to the Westbury Formation. *Lingula* sp. has not previously been described, but well preserved examples of a species of *Lingula* were recorded as being present at Garden Cliff, Westbury (Wang 1993). No other references to lingulids have been found in the literature and it is possible that the Garden Cliff occurrences and the species described here from the Dundry Hill (Elton Farm) borehole may be the only records.

Discinisca townshendi on the other hand has been recorded many times since its first description by Davidson (1851), but always as isolated occurrences at a few levels in the Westbury Formation. Its most notable occurrence is as quite large specimens in the matrix of the local conglomerates which bank up against an 'island' of Carboniferous Limestone in Vallis Vale near Frome, but it can also be found in the 'normal' dark grey shales of the Westbury Formation, as at Lavernock, near Penarth. The relatively large size of *D. townshendi*, which can reach 50 mm long, is rather surprising when compared with that of many inarticulate brachiopods and the general size of Penarth Group shelly fauna as a whole.

Specimens figured on Plate 23 are held in The Natural History Museum, London (figs 2–5), British Geological Survey, Keyworth (figs 1, 7–9) and the National Museum of Wales, Cardiff (fig. 6).

SPECIES DESCRIPTIONS

Phylum BRACHIOPODA Duméril, 1806
Class LINGULATA Gorjansky and Popov, 1985
Order LINGULIDA Waagen, 1885
Superfamily LINGULOIDEA Menke, 1828
Family LINGULIDAE Menke, 1828
Genus LINGULA Bruguière, 1797
Lingula sp.
Plate 23, figure 1

Description. Shell elongate, lateral margins subparallel, slightly convex. Surface covered with prominent, dark, concentric growth lines.
Remarks. The single specimen lacks the posteriormost portion of the shell and thus it is not possible to establish whether it is a pedicle or a brachial valve. Also, no internal structures can be seen. Therefore, no specific assignment can be made. However, general preservation is good and darkening related to original colour is present. The specimen is preserved in dark shale typical of the Westbury Formation.
Distribution. One record from the Westbury Formation at depth 191·9 m in the Dundry (Elton Farm) borehole, Somerset.

Order ACROTRETIDA Kuhn, 1949
Suborder ACROTRETIDINA Kuhn, 1949
Superfamily DISCINOIDEA Gray, 1840*b*
Family DISCINIDAE Gray, 1840*b*
Genus DISCINISCA Dall, 1871
Discinisca townshendi (Davidson, 1851)
Plate 23, figures 2–9

Description. Valves markedly dissimilar. Brachial valve large, sub-conical, smoothly convex. Pedical valve smaller, almost flat to gently concave, sub-circular. Pedicle track moderately broad, lanceolate to sub-oval, extending almost to posteror margin. Ornament of fine concentric growth rings, coarser and less regular on brachial valve.

Remarks. The foundation and earliest descriptions of *Discinisca townshendi*, as with so many taxa in the Penarth Group, are confused and insecure. It was first described by Davidson (1851), but he attributed the authorship to Forbes. This is invalid as Forbes' record was never published. Davidson (1851) applied the first published name, *Orbicula townshendi*, after its discoverer, based on specimens he initially believed to be from the Oxford Clay, but neither locality nor horizon were clear. In the Appendix to Volume 1 of his monograph Davidson (1855, p. 14) stated that the correct provenance was the Lias of Freherne Cliff near Newnham, Gloucestershire; he also used the name *Discina townshendi* in the Appendix. More light was shed on the origins of the original specimen in Davidson's supplement to his monograph (1876/78), where it was stated that there may have been confusion regarding the locality of the original find and that it may in fact have come from 'Rhaetic' levels at Garden Cliff.

The classification also evolved further in Davidson's supplement (1876/78) where the author claimed that the original species was synonymous with *Discina (Orbiculoidea) babeana*, a d'Orbigny species from levels in France considered to be equivalent in some part to the British 'Rhaetic'. Davidson (1876/78) further stated that the species had two new records, having been found in the 'Flinty-bed' at Beer Crowcombe (Moore 1861*b*) and also in the conglomeratic 'Rhaetic' rocks at Vallis Vale near Frome (Moore 1867). Both of these horizons equate with some portion of the Westbury Formation. Thus, the first undoubted 'Rhaetic' record of *Discinisca townshendi* in Britain was by Moore (1861*b*) and he also found comparable specimens elsewhere in the 'Rhaetic' (Moore 1867). Davidson's (1876/78) proposed synonymy with the species *babeana* was followed by Moore in a later paper (1881), and a systematic study using comparison of type material of *babeana* from France with the British material might be expected to confirm this synonymy. However, given the scope of this field guide it is thought preferable at this stage to maintain the use of *townshendi* as the species name, on the basis of its familiarity.

During the twentieth century with restriction of the use of the taxon *Discina* to Recent brachiopods (Rowell, *in* Williams *et al.* 1965), the name *Orbiculoidea townshendi* became established and has been used widely, but no taxonomic studies have been done since Davidson's original description in 1851 and his further remarks in 1876/78. A comparison of the descriptions of the genera *Orbiculoidea* and *Discinisca* in Rowell (*in* Williams *et al.* 1965), suggests that *townshendi* would be more appropriately assigned to *Discinisca*, mainly on the basis of the fairly broad suboval/lanceolate pedicle track which contrasts with the narrower one characteristic of *Orbiculoidea*.

Distribution. Rare. Only recorded from the Westbury Formation or its equivalents. Occurrences are concentrated in the south-west of the outcrop.

ANNELIDS

Annelid body fossils are extremely rare in the Penarth Group, which reflects the minimal preservation potential of their soft anatomy, but one species, *Archarenicola rhaetica* Horwood, 1912, has been described. The original description and detailed drawings of specimens from Leicestershire (Horwood 1912) are very clear, but the specimens themselves are not as well-preserved or unambiguous as the drawings suggest and the actual details, notwithstanding deterioration of the specimens since 1910 (the date of their discovery), are fairly obscure (Pl. 23, figs 12–14). It is possible that the remains are those of annulated burrow fills (i.e. trace fossils) as they are associated with other undoubted burrow fills, yet the latter are structureless and less regular in outline and there is little doubt that the segmented fossils are those of an annelid. Very few other records or specimens of *A. rhaetica* exist, but a specimen attributed to the genus *Archarenicola*, regrettably lacking details of locality or horizon, was located in the 'Rhaetic' collections of the Bristol City Museum and is illustrated on Plate 23, figure 15. Trace fossils attributable to the activities of annelids are common and give a more reliable indication of the actual status of annelids in the Penarth Group biota.

The minute jaw parts of certain polychaete worms ('bristle-worms') are chitinous and these elements, known as scolecodonts, are capable of preservation as fossils, and a small number have been isolated from acid insoluble residues of Langport Member limestones (Pl. 23, figs 10–11). They are characterized by a single smooth cusp with a non-denticulate blade occupying the lower half of the inner face. Such specimens may be assignable to *Glycera* sp. (see Szaniawski 1974). Indeterminate

PLATE 23

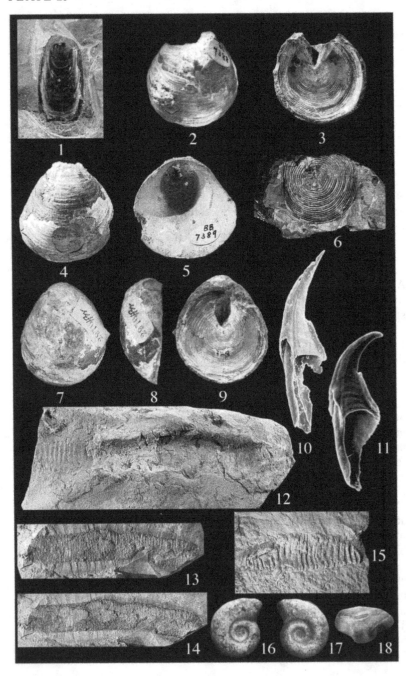

scolecodonts have also been identified in residues resulting from palynological preparations (e.g. Warrington 1978, 1981).

Specimens figured on Plate 23 are held in the New Walk Museum, Leicester (figs 12–14), Bristol City Museum and Art Gallery (fig. 15) and the A. Swift collection, Department of Geology, Leicester University (figs 10–11).

SPECIES DESCRIPTION

Phylum ANNELIDA Lamarck, 1809
Class POLYCHAETA Grube, 1850
Order DRILOMORPHA Hatschek, 1893
Family ARENICOLIDAE Audouin and Edwards, 1833
Genus ARCHARENICOLA Horwood, 1912
Archarenicola rhaetica Horwood, 1912
Plate 23, figures 12–15

Description. Cuticle bears annuli of two sizes, possibly forming segments. Cuticular surface of type specimen covered by epidermal papillae, paratype exhibits polygonal areas between somites. Paired appendages located on alternate annuli, consisting of capillary notopodial setae. Traces of frilled prostomium on head (based on Horwood 1912, p. 399).

EXPLANATION OF PLATE 23

cop.=copy of

Fig. 1. *Lingula* sp.; BGS Bx 2683; Westbury Formation; Dundry (Eldon Farm) Borehole, Somerset, depth 191·9 m; ×1·5.

Figs 2–9. *Discinisca townshendi* (Davidson, 1851); Westbury Formation. 2–5, 7–9, Hapsford Bridge, Vallis Vale, near Frome, Somerset. 2–3, BMNH BB 7388. 4–5, BMNH BB 7389. 7–9, BGS HBW 2970. 6, NMW 68·375·G1; 'cycle 3'; Lavernock Point, South Glamorgan. All ×1.

Figs 10–11. *Glycera*? sp.; Langport Member. 10, AS 6311/4; Dark Lane Copse Quarry, near Lighthorne, Warwickshire. 11, AS 32/36; Normanton Hills, near Loughborough, Leicestershire. Both ×100.

Figs 12–14. *Archarenicola rhaetica* Horwood, 1912; Westbury Formation. 12, paratype, NW 37'1960; Glen Parva, Leicester. 13 (oblique lighting) –14 (direct lighting), holotype, NW 32'1923; Spinney Hills, Leicester. Both x 0·75.

Fig. 15. *Archarenicola*? sp.; BCMAG 113/1976; 'Rhaetic'?; locality unknown; ×0·75.

Figs 16–17. Indeterminate ammonite; BCMAG Ce 9715; left (16) and right (17) sides; cop. Donovan *et al.* 1989, text-fig. 1A–B; Westbury Formation; Hampstead Farm Quarry, near Chipping Sodbury, Gloucestershire. Both ×5.

Fig. 18. *Chiton rhaeticus* Moore, 1861b; BRLSI M 35B; Westbury Formation; Vallis Vale, near Frome, Somerset; ×2.

Remarks. Horwood's (1912) paper in which he erected *A. rhaetica* gave extensive detail on each of the two specimens available to him, but today it is difficult to verify some of this due to deterioration of the specimens. In particular, traces of the appendages are very obscure. Some sort of coating compound, applied at some stage to protect the paratype (Pl. 23, fig. 12), also obscures detail.

There is confusion regarding which of the specimens is actually the type specimen. In the general notes on the specimens (Horwood 1912, pp. 396–399) Specimen A is stated to be represented by figure 2 on the plate and Specimen B by figure 1, yet in the systematic notes on p. 399 the reverse is given, i.e. A as figure 1 and B as figure 2. Horwood himself calls Specimen A the type. Measurement of the specimens reveals that Specimen B is in fact represented by figure 2 and thus Specimen A, figure 1, is the type. Unfortunately, pieces of Specimen A have become detached over the years and detail subsequently lost. The two specimens, despite both being attributed to *A. rhaetica*, are superficially quite dissimilar, but Horwood put this down to different degrees of compaction and the masking effect of matrix on the paratype.

The specimen from Bristol City Museum (Pl. 23, fig. 15) has been attributed at some stage to *Archarenicola* but in fact bears little resemblance to the original Horwood specimens beyond the possession of an annulated body. It may represent a different species, but the lack of horizon, locality details and corroborative specimens does not allow further interpretation.

Since Horwood's original discoveries and description, there have been no further descriptions of annelids from the Penarth Group, but a focused search in the Westbury Formation may yield further specimens.

Distribution. Westbury Formation, Glen Parva and Spinney Hills, Leicester.

CEPHALOPOD

Indeterminate ammonite
Plate 23, figures 16–17

Remarks. Cephalopods are almost unknown from the Penarth Group, the only record being a single very small ammonite from Hampstead Farm Quarry, near Chipping Sodbury, Gloucestershire [ST 726 839]. This was recovered from the uppermost bed of the Westbury Formation. The specimen is only 3·8 mm long and details are rather worn. It was reported by Donovan *et al.* (1989) and they concluded that it was of psiloceratid affinity, suggesting a commonality with Jurassic, rather than Triassic, forms. The presence of ammonites in lower Penarth Group beds, especially of Jurassic aspect, if proven, would indicate a radical reassessment of the age of the group, but whilst the single specimen stands alone, Donovan *et al.*'s (1989) cautious approach is followed here. More finds are clearly needed. The specimen is held in the collections of the Bristol City Museum and Art Gallery.

AMPHINEURAN

Plate 23, figure 18

Remarks. Members of the class Amphineura (Mollusca) are rarely recorded as fossils and in recent times have received little systematic study. Only one record has been published of examples from the Penarth Group (Moore 1861*b*); these specimens were recovered from the basal Westbury Formation clays at Vallis Vale, near Frome, Somerset. Moore named his species *Chiton rhaeticus* (Moore 1861*b*, p. 511, pl. 16, figs 28–29; herein Pl. 23, fig. 18). This material would now be allocated to the subclass Polyplacophora, otherwise known colloquially as 'chitons', but the lack of comparitive material and modern systematic studies precludes further evaluation of these specimens.

13. CONODONTS

by ANDREW SWIFT

Conodonts, long an enigmatic group of marine fossils, are now generally considered to be a group of extinct chordates whose affinites lie with the jawless fish (for a review of the conodont affinity debate see Aldridge and Purnell 1996). However, not all questions regarding their affinites and biology have been answered and they remain a distinctive and problematical group of organisms. The conodont fossil record is long, extending from at least the Middle Cambrian to the uppermost Triassic. Extremely rare fossils preserving traces of the soft bodies of conodonts show a sinuous eel-like morphology around 40 mm long (Carboniferous examples; Briggs *et al.* 1983) to an extrapolated 400 mm long (Ordovician examples; Gabbott *et al.* 1995). With the exception of one poorly preserved and incomplete specimen from the Silurian, no other conodont body fossils are known. Conodonts are almost entirely represented in the fossil record by the abundant remains of the only part of their anatomy normally preserved, their oral-pharyngeal food processing apparatus. This consisted of numbers of microscopic, phosphatic, tooth-like elements arranged in differentiated functional series. Within taxa, these elements are traditionally identified using a letter-based notation (P, M, S; further subdivided Pa, Pb, Sa, Sb, etc.) to indicate their position within the apparatus. New information on the architecture of conodont apparatuses suggests that a revision of this system is required (Purnell and Donohue 1998). Element length varies but only rarely exceeds 2–3 mm. Conodonts were an extremely successful group, demonstrating rapid diversification and evolution. There were many distinctive, short-lived species whose elements have been used to define biostratigraphical zones for almost the entire period of their existence.

Triassic conodonts have been recovered in Britain only from the Penarth Group (Swift 1989, 1995*b*). The Lower Triassic in Britain is developed in continental, restricted or non-marine facies with little or no potential for the recovery of conodont elements. Within the Penarth Group, only the 'White Lias' micritic facies of the Langport Member is known to contain conodont elements. These are very small (150–400 µm long), thin (one or two lamellae only) and almost transparent, making their detection difficult. Yet they are by no means rare in porcellanous splintery micrites in Warwickshire, Leicestershire and Nottinghamshire (Swift 1995*b*). Conodonts are present, but less common, in the the Langport Member as developed in south-west Britain, and, despite the processing of a number of samples, have yet to be recovered from the Westbury Formation. The reasons for this are unknown, but further investigations may yet reveal their presence. The largely non-marine environments prevalent during the deposition of the Cotham Member were not conducive to colonization by conodonts, and the few more open marine levels in south-west Britain have so far proved barren.

The conodonts of the Langport Member flourished at a critical period in conodont evolution, very close to the extinction of the group as currently conceived. No records are known from rocks of undisputed Jurassic age, although a few elements have been recovered from the Rhaetian portion of the Lias Group (i.e. basal beds lacking psiloceratid ammonites) in Britain (Swift 1989, 1995*b*). However, the possibility of reworking cannot be discounted for these elements. Penarth Group conodonts appear to show modification in the composition of their apparatuses as well as in size, and none has a 'normal' plan when compared with the more familiar apparatuses of conodonts from earlier geological systems i.e. typically containing six or more distinct morphotypes. Evidence from British collections suggests that *Misikella coniformis* had just one type of element in its apparatus whilst *Chirodella verecunda* contained only two. Other elements, which are tentatively assigned to species of *Prioniodina*, are very few in number and their natural species associations cannot be meaningfully analysed, but, despite sharing similar denticulation, they do not appear to fit into established apparatus templates. Indeed, at least ten out of a total of 19 identifiable specimens are different (five of these are described here and figured on Pl. 24, figs 6–10). The reasons for these aberrations are not clear, but it is possible that the Penarth Group conodont populations were under some form of stress which prompted biological modifications, perhaps the very environmental and/or biological pressures which in a short space of time led to their extinction.

Systematic notes are based on the extended descriptions of Swift (1995*b*, pp. 51–58), and the reader is referred to this work for further information. All figured specimens are held in the micropalaeontology collections in the Department of Geology at the University of Leicester.

SPECIES DESCRIPTIONS

Phylum CHORDATA Bateson, 1886
Class CONODONTA Pander, 1856
Order PRIONIODINA Sweet, 1988
Family GONDOLELLIDAE Lindström, 1970
Genus MISIKELLA Kozur and Mock, 1974.
Misikella coniformis Swift, 1995*b*
Plate 24, figure 1

Description. Apparatus unimembrate, consisting of a coniform Pa element. Cusp high, wide, laterally compressed and curved posteriorly. Basal cavity deep, flared, occupying entire underside.

Remarks. Species of *Misikella* are characteristic of the upper part of the Triassic and are amongst the last conodonts in the fossil record. Only two elements of *M. coniformis* have been found in the Penarth Group, out of a total of over 1000 specimens now in collections, but the species is relatively common in the Rhaetian of Cyprus (Fåhræus and Ryley 1989) and has also been recovered from the Rhaetian of Hungary (Kozur and Mock 1991). Although reduced to a single denticle, the relationship with Pa elements from other *Misikella* species is evident in the lateral compression of the cusp and fully excavated underside.

Distribution. Langport Member. Normanton Hills, near Loughborough, Leicestershire and near Cotgrave, Nottinghamshire.

Family *Incertae sedis*
Genus CHIRODELLA Hirschmann, 1959
Chirodella verecunda Swift, 1995*b*
Plate 24, figures 2–5

Description. Apparatus bi-membrate. Both elements characterized by small size, delicacy and thinness. Denticles very numerous, needle-like. M element (Pl. 24, figs 2–3) digyrate, strongly arched, paired, asymmetrical. Cusp and surrounding denticles twisted and incurved. Outer lateral process long, downwardly directed, gently curved. Denticles inclined tightly to process, sometimes curved posteriorly. Inner lateral process short, carrying two to four denticles inclined posteriorly. Basal opening narrow, flared moderately beneath cusp. Sc element (Pl. 24, figs 4–5) bipennate, paired, highly compressed. Anterior process short, directed downwards, bearing one to three, most often two, diverging denticles. Posterior process long, bearing numerous straight denticles which become more inclined distally. Cusp fairly high, inclined posteriorly. Basal opening very narrow. Aboral margin curved, most acutely beneath cusp. Costae often developed on denticles; also on anterior edge of anterior process where they may become thickened.

Remarks. *Chirodella verecunda* is by far the most common conodont species in the Penarth Group and its elements account for around 97 per cent. of all specimens recovered. An analysis of the 800+ elements assignable to this species reveals a ratio of approximately one M element to 15 Sc elements. The two elements are of familiar general morphology and are similar to many other M and Sc elements from the conodont record, but no other comparable bi-membrate species are known. Ryley and Fåhræus (1994) described similar species of *Chirodella* from the Upper Triassic of Cyprus, but while their species apparatuses consist, like *C. verecunda*, of digyrate ('M') and bipennate ('Sc') elements, Ryley and

Fahraeus (1994) recognized more morphotypes within their basic apparatus plans, as well as more variation.

It might be speculated that the apparatus of *C. verecunda* is a relict of a more complex conventional apparatus which has been reduced both in element number and size, and represents an advanced stage in the complete loss of elements as a result of biological modification forced by environmental and/or biological stress.

Distribution. Langport Member throughout its outcrop. A few elements have been recovered from the basal Rhaetian horizons of the Lias Group in Nottinghamshire (Swift 1989, 1995*b*).

Family PRIONIODINIDAE Ulrich and Bassler, 1926
Genus PRIONIODINA Ulrich and Bassler, 1926?

Remarks. The elements tentatively assigned to *Prioniodina* spp. are diverse, but very rare, and have been recovered mainly from localities in Leicestershire and Nottinghamshire. All are blade-like and possess widely spaced discrete denticles with crimped edges. They may be members of species of a new genus, but the low numbers recovered do not allow reconstruction or discrimination of species. It is also difficult to refer these elements to the usual positional notation scheme. Although all appear robust, they are, like all Penarth Group conodont elements, extremely small and fragile, and most are broken.

Prioniodina? spp.
M element
Plate 24, figure 6

Description. Slightly asymmetrical digyrate element, strongly arched. Both processes broken; outer? lateral more complete, short, slightly twisted, bearing one high, erect denticle. Inner? lateral process thicker at break, probably originally much longer than outer? lateral. Cusp high, erect, broad. Basal cavity moderately flared beneath cusp, extended as groove along posterior process; developed only for very short distance along anterior process.

Remarks. The element probably occupied an M position, but differs somewhat from the usual morphology. It is more symmetrical and the two denticles either side of the cusp are comparable in size and shape and are erect like the cusp, in the manner of an Sa element. However, the basal cavity, although not so large as in most M elements, is moderately, but clearly, asymmetrical. In addition, the outer? lateral process is slightly twisted.

Distribution. One specimen from the Langport Member at Normanton Hills, near Loughborough, Leicestershire.

P? element type A
Plate 24, figure 7

Description. Fairly short, straight, bearing seven erect, widely spaced denticles with crimped edges. Denticles high at one end, diminishing evenly in height, terminating in very short upturned margin. Basal opening very narrow, faintly defined.

Remarks. Elements of this general morphology are not uncommon in the Middle–Upper Triassic, but none of the published forms has both the widely spaced denticles and the regular decrease in denticle height from one end of the bar to the other which is characteristic of the British specimen. It is not possible to refine further the position this element occupied within the apparatus (i.e. Pa, Pb, Pc, etc.). It bears some resemblance to S elements from apparatuses of gondolellid species from the Permian (Rieber 1980).

Distribution. Langport Member. One specimen from Normanton Hills, near Loughborough, Leicestershire.

P? element type B
Plate 24, figure 8

Description. Small, with three processes. Lateral process broken at junction with cusp; posterior process broken after first denticle. Anterior process downwardly directed, bearing two widely spaced denticles. Cusp strongly developed, inclined posteriorly, bearing a prominent costa which continues the line of junction with the lateral process. Basal opening very narrow, extending a short distance along all processes from greatest development beneath cusp.

Remarks. The single specimen is broken and its affinities cannot be accurately evaluated, but it seems most likely to have occupied a P position in its apparatus. P elements with three processes are not uncommon in earlier genera e.g. *Prioniodus* Pander in the Ordovician and *Kladognathus* Rexroad in the Carboniferous. Alternatively, it may have occupied an S position, as such elements also often bear three processes.

Distribution. Langport Member. One specimen from Normanton Hills, near Loughborough, Leicestershire.

Sb element
Plate 24, figure 9

Description. Small, moderately arched, bipennate. Anterior process fairly short, bearing two semi-erect, widely spaced, crimped denticles. Posterior process longer, broken, bearing three similar denticles. Cusp high, slightly inclined and curved towards posterior. Basal cavity slightly flared beneath cusp, extended a little way along processes.

PLATE 24

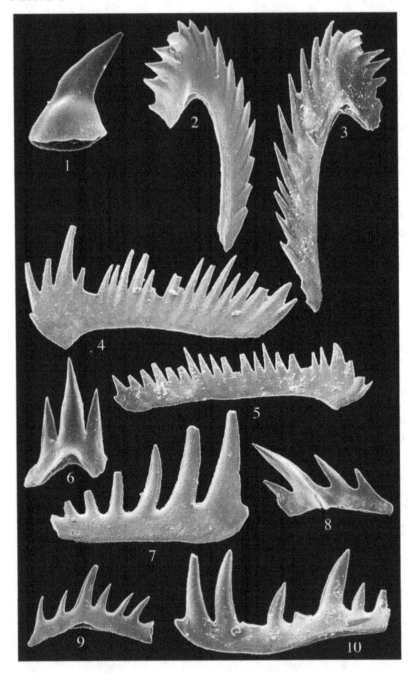

Remarks. This element is similar to Pb types in general morphology, but the small, narrow basal cavity and overall delicacy is taken to indicate that it occupied an Sb position.

Distribution. Two specimens from the Langport Member. Normanton Hills, near Loughborough, Leicestershire.

Sc element
Plate 24, figure 10

Description. Long, bipennate, with sinuous aboral margin. Anterior process more complete, long, curved upwards, bearing four or five discrete, widely spaced denticles. Posterior process broken near cusp, probably originally very long. Cusp moderately high, inclined posteriorly. Other anterior denticles of comparable height to cusp. Basal cavity slightly flared beneath cusp, extended as groove along anterior process as far as first denticle from cusp.

Remarks. A very distinctive element and, of all the elements referred to *Prioniodina*?, is least similar to published forms. Allocation to an Sc position is made on the basis of the lack of arching and the upswept anterior process (typical of many 'hindeodellid' elements). Additionally, by extrapolating the length of the broken posterior process on the basis of

---

EXPLANATION OF PLATE 24

All specimens from the Langport Member, cop.=copy of

Fig. 1. *Misikella coniformis* Swift, 1995*b*; AS L6/1B; near Cotgrave, Nottinghamshire; cop. Swift 1995*b*, pl. 5, fig. 3.

Figs 2–5. *Chirodella verecunda* Swift, 1995*b*; 2, dextral M element; AS 6215/6; Gotham Hills, Nottinghamshire; cop. Swift 1995*b*, pl. 6, fig. 9. 3, sinistral M element; AS 6318/10; near Clipston, Nottinghamshire; cop. Swift 1995*b*, pl. 6, fig. 11. 4, dextral Sc element; AS 6218/20; Gotham Hills, Nottinghamshire. 5, sinistral Sc element; AS 6315/11; Normanton Hills, near Loughborough, Leicestershire.

Fig. 6–10. *Prioniodina*? spp.; 6, M element; AS L5/16; Normanton Hills, near Loughborough, Leicestershire; cop. Swift 1995*b*, pl. 5, fig. 9. 7, P? element type A; AS 6234/15; Normanton Hills, near Loughborough, Leicestershire; cop. Swift 1989, pl. 37, fig. 14. 8, sinistral P? element type B; AS 6246/11; Normanton Hills, near Loughborough, Leicestershire; cop. Swift 1989, pl. 37, fig. 12. 9, dextral Sb element; AS 6234/14; Normanton Hills, near Loughborough, Leicestershire; cop. Swift 1989, pl. 37, fig. 10. 10, dextral Sc element; AS 6314/7; Normanton Hills, near Loughborough, Leicestershire; cop. Swift 1995*b*, pl. 5, fig. 13.

All lateral views, ×200.

the thickness of the process at the break, it is postulated that the element was originally very long, a feature typical of many Sc elements. Three elements definitely assignable to this type have been recovered, and some fragments may be the remains of other comparable elements, making this the most common of the morphotypes referred to *Prioniodina*?.

Distribution. Langport Member. One specimen from Normanton Hills, near Loughborough, Leicestershire; one specimen from near Clipstone, Nottinghamshire; one specimen from a loose block from Hathern, Loughborough, Leicestershire.

14. FISH

by CHRISTOPHER J. DUFFIN

The majority of fish fossils from the Penarth Group are disarticulated, dissociated components such as scales, teeth, fragments of jaws, fin rays, dorsal fin spines, cephalic spines and robust bones such as the quadrates, parasphenoids and tooth-bearing elements of the pterygoid complex. Most specimens have been collected from the bone beds of the Westbury Formation; the black shales, which form the bulk of the Westbury Formation, contain only occasional isolated fragments. Some other specimens originate from the mixed sediments which are found in fissures within the Carboniferous Limestone sequence of south-west England and South Wales. These atypical sediments are difficult to date, and encompass an age range thought to be Late Triassic–Early Jurassic. However, by comparison with established Rhaetian faunas elsewhere, the taxa described here are believed to come from levels broadly equivalent to some part of the Penarth Group or, at least, to be of Rhaetian age.

Very few articulated fish are known. Exceptions are an undescribed partial hybodont shark skull from Watchet, some tiny specimens of *Legnonotus cothamensis* and *Pholidophorus higginsi* from the Cotham Member of Aust Cliff (Egerton 1854) and a partial large lungfish probably from the Penarth Group at Blue Anchor Point (Pl. 25).

Teeth and scales can be recovered from calcareous bone bed horizons using acetic or formic acid digestion, which will not affect the phosphatic component. Other non-calcareous horizons may break down when treated with hydrogen peroxide or petroleum ether.

Figured specimens are held in The Natural History Museum, London (BMNH prefixes), Bath Royal Literary and Scientific Institute (BRLSI prefixes), Bristol City Museum and Art Gallery (BCMAG prefix), New Walk Museum, Leicester (NW prefix), School of Earth Sciences, University of Portsmouth (OUP prefix), A. Swift collection, Department of Geology, University of Leicester (AS prefix) and the CJD personal collection.

SPECIES DESCRIPTIONS

Superclass CHONDRICHTHYES Huxley, 1880

Remarks. Chondrichthyes includes the sharks, rays and chimaeras, all of which have cartilaginous skeletons. Cartilage is usually non-mineralized and as such has a low preservation potential – the high protein and other organic matter content causes most skeletons to be lost through

PLATE 25

Fig. 1. *Ceratodus*? sp.; OUP 98/35; loose block; Blue Anchor Point, Somerset; partially articulated incomplete skeleton; ×0·25.

decomposition. Cartilage may be calcified in tiny prisms, which are sometimes robust enough to survive in bone bed deposits. The scales of chondrichthyans are of both simple growing and non-growing types. Males develop pelvic claspers supported by specialized cartilages serving as intromittent organs for the efficient deposition of sperm into the female reproductive tract. The dorsal fins are primitively supported anteriorly by fin spines.

The group is divided into the elasmobranchs or selachians (sharks, skates and rays) and holocephalans (chimaeras or rabbit fishes). These differ considerably in the form and mode of growth of their teeth, as well as in their life habits.

Class ELASMOBRANCHII Bonaparte, 1838
Cohort EUSELACHII Hay, 1902
Superfamily HYBODONTOIDEA Owen, 1846
Family HYBODONTIDAE Owen, 1846

Remarks. Hybodont sharks are commonly represented in marine Mesozoic rocks by isolated teeth, scales, dorsal fin spines and cephalic spines. The latter are paired structures confined to male individuals and are presumed to have been important during mating. The recurved, enamelled spine was anchored into the superficial tissues by a trifid base. Articulated, whole-bodied and partial specimens of hybodont sharks from the Lower Jurassic of Lyme Regis, Holzmaden (Germany) and the Lower Cretaceous of Sussex and the Isle of Wight indicate that the cephalic spines were situated just behind the orbits or even further back above the braincase. Hybodont sharks possessed fairly densely calcified cartilages, scales that were capable of growing by odontode (dentine unit) fusion and teeth with single crystallite enameloid. Furthermore, the teeth show anaulacorhize vascularization of the root, i.e. the root is perforated by numerous tiny vascular foraminae with no preferred spatial organization over the root surfaces.

A reconstruction of a typical Jurassic hybodont shark is given in Text-figure 19A.

Genus 'HYBODUS' Agassiz, 1837
'*Hybodus*' *minor*, Agassiz, 1837
Plate 26, figures 1–4

Description. Species known from isolated teeth and dorsal fin spines only.

Teeth (Pl. 26, figs 1–2) up to 10 mm high, more usually around 3 mm, symmetrical, with high, upright central cusp flanked by up to four pairs of lateral cusplets. Central cusp has fairly wide base, lingually inclined in lateral view. Fairly coarse vertical ridges descend cusps from apices,

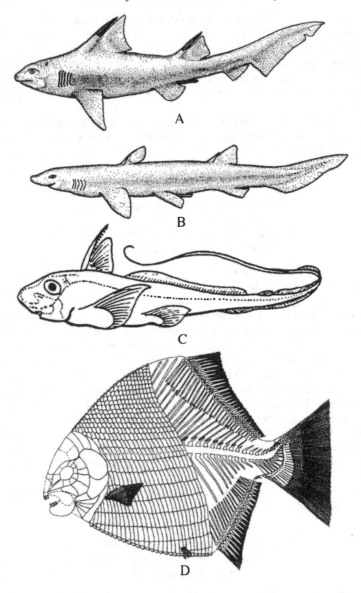

TEXT-FIG. 19. Reconstructions of Penarth Group fish. A, hybodont shark (after Maisey 1982); B, palaeospinacid shark (after Duffin and Ward 1993, p. 79, fig. 13); C, Recent holocephalan (chimaeran) shark (after Janvier 1996, fig. 3·15A); D, a Triassic bony fish, *Sargodon tomicus* (after Tintori 1982, p. 421, fig. 1). Not to scale.

occasionally bifurcating basally. Cusps have moderately sharp lateral cutting edges. Root shallow, labial face less than one-fifth of total tooth height. Remainder of root projects lingually in a so-called 'lingual torus', roughly semicircular in basal view and perforated by numerous vascular foramina.

Fin spine (Pl. 26, figs 3–4) up to 150 mm long, gently recurved toward tip with oval cross section. Lateral walls ornamented by series of non-branching longitudinal costae or ribs. The non-costate basal part of the spine was embedded in tissues of the back. Posterior wall open toward base, allowing articulation with basal cartilage of dorsal fin, closed for upper or distal two-thirds of spine length. Posterior spine wall has single row of reflexed denticles situated medially.

Remarks. '*Hybodus*' *minor* was originally defined by Agassiz (1837) on the basis of a small dorsal fin spine from a Westbury Formation bone bed ('Rhaetic Bone Bed') at Purton Passage on the south bank of the River Severn, near Berkeley, Gloucestershire. He subsequently (Agassiz 1837) ascribed teeth to the same taxon, although it is by no means certain that the teeth and fin spines belong together (see discussion below).

The fin spines of *Hybodus* are indistinguishable from those of *Acrodus*, *Lissodus* and *Polyacrodus*. Correctly, the species name '*Hybodus*' *minor* should be applied only to fin spines resembling that originally described by Agassiz, and as such '*Hybodus*' *minor* is a *nomen dubium*, because of the lack of diagnostic characters. Furthermore, the hybodont sharks '*Hybodus*' *cloacinus* and *Lissodus minimus* described below presumably also possessed dorsal fin spines. Allocation of hybodont fin spines to individual species within the Penarth Group is attended with great uncertainty, although very tentative assignments may be made on the basis of size (with those of '*Hybodus*' *cloacinus* amongst the largest examples).

Other dorsal fin spines which have been described from the Penarth Group include *Hybodus laeviusculus* Agassiz, *Hybodus punctatus* Davis, and *Hybodus austiensis* Davis. These species are impossible to distinguish from the fin spines of '*Hybodus*' *minor* and are considered to be junior synonyms of this species.

Similar problems surround the teeth described by Agassiz (1837) and ascribed by him to '*Hybodus*' *minor*. The shallow, lingually offset root is reminiscent of that in certain primitive neoselachians ('modern level' sharks) and very different from the more tabular root seen in the teeth of *Hybodus*. Indeed, the teeth of '*Hybodus*' *minor* are very similar to those recently described for *Rhomphaiodon nicolensis* from the Norian of France and Germany (Duffin 1993a). *Rhomphaiodon* teeth have a distinctive triple-layered enameloid indicating neoselachian affinities. Also, the crowns of '*Hybodus*' *minor* are similar to those of the Triassic neoselachian *Synechodus* (see below).

Thus, it is clear that a degree of taxonomic confusion was created at the time of the erection of '*Hybodus*' *minor* and this has been compounded by the continued indiscriminate use of the species name. Formal taxonomic revision is obviously required. Until that work is completed both spines and teeth are allocated for convenience as originally suggested by Agassiz (1837), and in accordance with traditional practice.

Distribution. '*Hybodus*' *minor* is well known throughout the European Rhaetian and has also been recorded from the Norian of Luxembourg (Duffin 1993*b*). Pre-Norian records cited in older literature suggest a Ladinian to Rhaetian range for the species, although corroboration is still needed (Duffin and Delsate 1993). Within the Penarth Group the teeth are ubiquitous, particularly in the Westbury Formation and, specifically, its bone-beds.

'*Hybodus*' *cloacinus* Quenstedt, 1858
Plate 26, figure 5

Description. Slender, elongate teeth up to 25 mm long are the only identified parts of this fish. Crown multicuspid with moderately high central cusp flanked by up to four pairs of fairly low, stubby lateral cusplets. Vertical ridges ascend crown from level of crown shoulder and may reach cusp apices. Ridges may bifurcate basally. Basal ridge of central cusp expanded to form labial node. Similar nodes may also be developed at labial bases of lateral cusplets. A tabular root, subequal to height of crown and projecting slightly lingually from crown underside, is seldom preserved.

Remarks. '*Hybodus*' *cloacinus* was originally described from the German Rhaetian around Tübingen, southern Germany. It has subsequently been recorded from most European Rhaetian sites, where it is always a fairly rare faunal component. There is a single record from the Lower Jurassic of Lyme Regis (Duffin 1993*d*), giving the species a range of Rhaetian to Sinemurian.

Distribution. Found rarely, but occurs throughout the Penarth Group.

Family POLYACRODONTIDAE Glückman, 1964
Genus PALAEOBATES von Meyer, 1849
Palaeobates reticulatus Duffin, 1998*b*
Plate 26, figure 6

Description. Species known from teeth only. These are slim, elongate and reach up to 6 mm long, 1 mm in diameter and 1 mm high. Robust, non-cuspidate crown ornamented by complex reticulate pattern formed by ridge bifurcation and anastomosis. Ridges commence half way up crown shoulder and coarsen towards crown midpoint. Median longitudinal groove may run length of crown.

Remarks. The teeth of *Palaeobates* are defined partly on morphology i.e. the low coronal profile, and partly on histology i.e. the crown is composed mainly of pallial dentine. The genus is exclusively Triassic, ranging from the Scythian to this, the latest record.

Distribution. Known only from Rhaetian fissure fills in the Carboniferous Limestone at the type locality of Holwell, Somerset.

Genus POLYACRODUS Jaekel, 1889
Polyacrodus holwellensis Duffin, 1998*b*
Plate 26, figure 7

Description. Known only from narrow, pointed teeth up to 4 mm long. Central cusp low, inclined distally, no flanking lateral cusplets. Non-branching vertical ridges ascend crown from crown shoulder both labially and lingually. Root subequal to crown in height, minimally offset from crown underside, with flat basal face and vascular foramina arranged in simple longitudinal rows labially and lingually.

Remarks. *Polyacrodus* is a poorly defined taxon which ranges from the Lower Permian to the Upper Cretaceous. Jaekel (1889) originally distinguished the genus from teeth of *Hybodus* on histological grounds. *Polyacrodus* teeth have crowns composed of dentine whereas crowns of *Hybodus* teeth are made of osteodentine.

The first appearance of this taxon may prove useful in correlation over a wide geographical area.

Distribution. Originally described from fissure fills at Holwell Quarry, Somerset and the base of the Cotham Member at Chilcompton, Somerset (Duffin 1998*b*).

Family LONCHIDIIDAE Herman, 1975
Genus LISSODUS Brough, 1935

Remarks. Sharks of the genus *Lissodus* were originally described on the basis of small articulated specimens from the Lower Triassic of South Africa (Brough 1935). They are typically hybodont in possessing stud-like dermal scales, cephalic spines on the skull behind the eyes in males, and two vertically ribbed dorsal fin spines supporting the median fins. Circumstantial evidence (i.e. co-occurrence) from the British Wealden (Lower Cretaceous) and North American Cretaceous suggests that the fin spines of *Lissodus* may have possessed a series of down-turned denticles arranged in a single row in the middle of the convex posterior wall of the dorsal fin spine. Also, material from the Lower Triassic of Angola suggests that the

PLATE 26

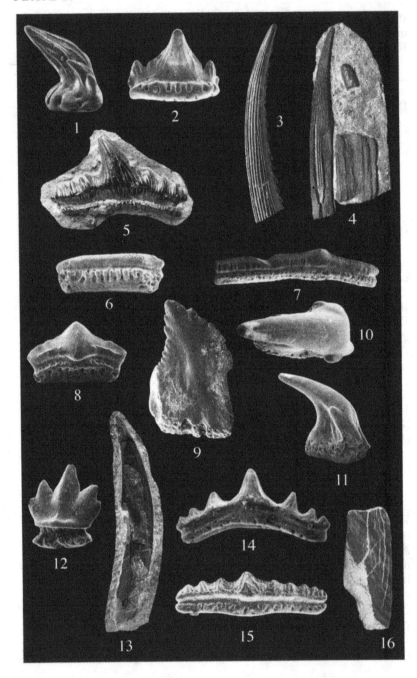

cephalic spine of *Lissodus* may be characterized by a 'convict-arrow' shape to the three-pronged base.

The teeth of *Lissodus africanus*, the type species, were not described in detail until recently (Duffin 1985). This study demonstrated that many species previously allocated to other genera, particularly *Acrodus*, belong in *Lissodus*. *Lissodus* ranges from the Upper Devonian through to the Upper Cretaceous of Europe, Africa and North America.

Lissodus minimus (Agassiz, 1839)
Plate 26, figure 8; Text-figure 20A–D

Description. Species established solely on the basis of isolated teeth up to 7 mm long. Crown low profiled, with stubby central cusp flanked by up to five pairs of very low lateral cusplets. Crown ornamented by series of often bifurcating vertical ridges which descend from cusp apices on labial

---

EXPLANATION OF PLATE 26

Figs 1–4. *'Hybodus' minor* Agassiz, 1837. 1–2, tooth; BRLSI CD 27; fissure infill =?Westbury Formation; Holwell, Somerset; lateral and lingual views; ×10. 3, dorsal fin spine; BMNH P 51443; 'Upper Triassic, Rhaetic Beds'; Barnstone, Nottinghamshire; lateral view; ×1. 4, dorsal fin spine; BMNH P 2778; basal Westbury Formation bone bed; Aust Cliff, Gloucestershire; lateral view; ×0·25.

Fig. 5. *'Hybodus' cloacinus* Quenstedt, 1858; tooth; BMNH P 45802; basal Westbury Formation bone bed; Garden Cliff, Westbury-on-Severn, Gloucestershire; lingual view; ×2.

Fig. 6. *Palaeobates reticulatus* Duffin, 1998b; tooth; holotype, BRLSI CD 18; fissure infill=?Westbury Formation; Holwell, Somerset; labial view; ×10.

Fig. 7. *Polyacrodus holwellensis* Duffin, 1998b; tooth; holotype, BRLSI CD 16; fissure infill=?Westbury Formation; Holwell, Somerset; labial view; ×10.

Fig. 8. *Lissodus minimus* (Agassiz, 1839); tooth; BRLSI CD 46; fissure infill=?Westbury Formation; Holwell, Somerset; labial view; ×3·5.

Figs 9–11. *Pseudodalatias barnstonensis* (Sykes, 1971). 9, lower lateral tooth; BRLSI CD 17; fissure infill=?Westbury Formation; Holwell, Somerset; lingual view; ×8. 10–11, upper tooth; BRLSI CD 45; fissure infill=?Westbury Formation; Holwell, Somerset. 10, occlusal view; 11, lateral view. Both ×18.

Fig. 12. *Vallisia coppi* Duffin, 1982b; tooth; holotype, BCMAG Cc 400; Westbury Formation; Vallis Vale, Frome, Somerset; labial view; ×10.

Figs 13–16. *Synechodus rhaeticus* Duffin, 1982a. 13, dorsal fin spine; holotype, NW OS 2'1978; basal Westbury Formation bone bed; Aust Cliff, Gloucestershire; lateral view; ×0·75. 14, anterolateral tooth; BRLSI M 181b; fissure infill=?Westbury Formation; Holwell, Somerset; lateral view; ×10. 15, extreme posterolateral tooth; BRLSI M181c; fissure infill=?Westbury Formation; Holwell, Somerset; lingual view; ×10. 16, fin spine fragment; BRLSI CD 20; fissure infill=?Westbury Formation; Holwell, Somerset; lateral view; ×2.

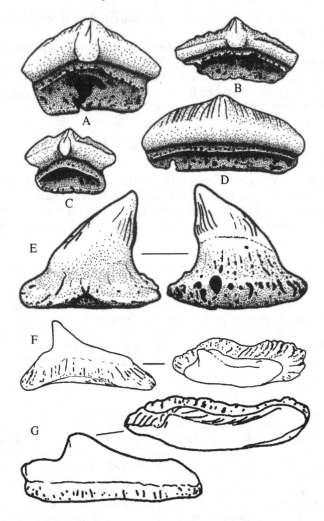

TEXT-FIG. 20. A–D, *Lissodus minimus* (Agassiz, 1839); fissure infill=?Westbury Formation; Holwell, Somerset. A, lateral tooth; BRLSI CD 57. B, anterolateral tooth; BRLSI CD 56. C, anterior tooth; BRLSI CD 55. D, posterolateral tooth, BRLSI CD 58. All in labial view; ×8 (after Duffin 1985, p. 124, fig. 14a-d). E–G, *Pseudocetorhinus pickfordi* (Duffin, 1998a); fissure infill=?Westbury Formation; Holwell, Somerset. E, labial and lingual views of anterior tooth; BRLSI CD 1997/1 (after Duffin 1998a, p. 161, fig. 3a-b). F, labial and occlusal views of anterior tooth; BRLSI CD 1997/8 (after Duffin 1998a, p. 167, fig. 6c, f). G, labial and occlusal views of posterolateral tooth; BRLSI CD 1997/10 (after Duffin 1998a, p. 168, fig. 7f-g). All ×20.

and lingual faces of crown. A longitudinal ridge surrounds tooth along surface of crown shoulder. Lateral margins of crown extend well beyond crown/root junction. Central cusp has distinctive peg-like expansion situated low down on labial side. Pressure scar resulting from tooth to tooth contact in the jaw often developed in corresponding position on lingual side of central cusp. Root approximately same height as crown and projects slightly lingually from crown undersurface. Root has very shallow upper face on labial side, bearing a longitudinal row of tiny vascular foraminae. Labial face of root concave in lower portion with much larger, randomly distributed foraminae.

Remarks. This species was originally placed in the genus *Acrodus* and appears cited as *Acrodus minimus* in most of the older literature. It has only recently been allocated to the genus *Lissodus* (Duffin 1985) and is one of the most abundant, best known members of the genus. The low-crowned teeth were well suited to crushing benthic shelled invertebrates such as bivalves and crustaceans.

Distribution. *Lissodus minimus* has been recorded from virtually all Penarth Group and European Rhaetian localities, where it forms one of the main components of the vertebrate fauna.

Family PSEUDODALATIIDAE Reif, 1978
Genus PSEUDODALATIAS Reif, 1978
Pseudodalatias barnstonensis (Sykes, 1971)
Plate 26, figures 9–11

Description. Teeth of lower jaw blade-like (Pl. 26, fig. 9), compressed labio-lingually, showing strong variation along length of jaw. Symphyseal teeth (located at junction of two lower jaws) symmetrical with triangular crown. Lateral cutting edges of all teeth strongly serrated. Root less than one-quarter of total tooth height, with two small vascular foramina entering the tissues close to base. Laterally, there is an articular facet on each side for the reception of subsequent teeth in the arcade. Successive lateral teeth progressively more asymmetrical and low crowned. Flattened, serrated cusp becomes progressively more posteriorly inclined.

Upper teeth (Pl. 26, figs 10–11) have prominent central cusp with almost circular cross section flanked by pair of very low lateral cusplets. Bifid root strongly divided, highly vascularized, with one prominent entrant foramen located at apex of inverted V-shaped basal face.

Remarks. The teeth of this species were originally described from the Westbury Formation formerly exposed in a disused railway cutting at Barnstone, Nottinghamshire (Sykes 1971), and subsequently recorded from a wide range of Penarth Group localities (Sykes 1974*a*). The dentition shows strong dignathic heterodonty – the needle-like upper teeth contrast strongly in morphology with the blade-like, labio-lingually flattened lower teeth. These morphologies closely parallel those of modern squalomorphic sharks such as *Scymnodalatias licha*, the Cookie-cutter shark. Like its modern counterparts, *Pseudodalatias* must have shed the teeth in its lower dentition as a complete articulated strip, whose integrity was maintained by close, complex tooth-to-tooth articulations. This is indicated by the fact that complete lower dentitional tooth strips have been found in the Norian deposits at Cene in Italy (Tintori 1980). Isolated teeth have also been found in the Rhaetian of Belgium (Duffin *et al.* 1983; Duffin and Delsate 1993) and France (Cuny *et al.* 1994).

The affinities of *Pseudodalatias barnstonensis* have been the subject of some debate. The morphology suggests that it was a squalomorphic shark, but work on the ultrastructure of the enameloid shows a unique structure more closely allied to that of hybodont sharks. Much more work on shark enameloid ultrastructure is required before the matter can be resolved.

Distribution. Widespread, but a rare component of the Penarth Group ichthyofauna.

Subcohort NEOSELACHII Compagno, 1977
Order and Family *Incertae sedis*
Genus VALLISIA Duffin, 1982*b*
Vallisia coppi Duffin, 1982*b*
Plate 26, figure 12

Description. *Vallisia* is known only from isolated teeth up to 2·5 mm high. Crown unornamented, upright to slightly inclined, with flattened labial and lingual faces. Occlusal margin multicuspid with higher central cusp. Crown base expanded to form lip which overhangs moderately deeply incised crown/root junction around entire tooth. Root forms about one-third of total height, base flat to convex with slightly flared margins. Basal surface of root crossed by one or two open or partially roofed median canals, with a vertically directed vascular foramen located in their floor.

Remarks. *Vallisia* is one of a number of early neoselachians which are difficult to place in extant orders.

Distribution. *Vallisia coppi* has so far been recorded from only two sites exposing the Penarth Group or its equivalent: Vallis Vale (the type locality) and the Holwell fissure fillings, both in Somerset. It is also known from the Belgian Rhaetian (Duffin *et al.* 1983).

Superorder SQUALOMORPHII Compagno, 1973
Order SYNECHODONTIFORMES Duffin and Ward, 1993
Family PALAEOSPINACIDAE Regan, 1906

Remarks. Palaeospinacid sharks are best known from partial and complete articulated specimens from the Lower Jurassic of Lyme Regis,

Holzmaden in southern Germany, and the Upper Jurassic lithographic limestones of Nusplingen, near Stuttgart (Duffin 1993*a*; Duffin and Ward 1993). They range from the Lower Triassic to Paleocene, and were small sharks somewhat resembling modern dogfishes (Text-fig. 19B). The form of the scales from articulated specimens suggests that they were fairly slow swimming bottom hunters of nearshore coastal waters. The two dorsal fins were each supported by a dorsal fin spine.

A review of *Synechodus* and its allies (Duffin and Ward 1993), concluded that the extensively employed name *Palaeospinax* is only applicable to a single specimen on which no diagnostic features can be found. This has resulted in the unsatisfactory, but not unique, situation where the valid family name is based upon a genus which cannot be adequately defined.

Genus SYNECHODUS A. S. Woodward, 1888
Synechodus rhaeticus (Duffin, 1982*a*)
Plate 26, figures 13–16

Description. Dorsal fin spine (Pl. 26, figs 13, 16) up to 75 mm long, gently recurved with approximately triangular cross section. Outer surface enamelled for around two-thirds of length, with exception of posterior wall. Enamel along anterior margin may be lost through in-life wear. Growth lines clearly visible in enamel, sweeping upward and backward from anterior margin. A rough, double row of isolated enamel tubercles up to 3 mm across, separated from the main body, located below basal margin of enamelled covering.

Teeth (Pl. 26, figs 14–15) multicuspid with higher central cusp flanked by up to four pairs of lateral cusplets reducing in size away from central cusp. Cusps upright to slightly inclined lingually, ornamented by series of vertical ridges which descend from cusp apices on both labial and lingual faces of crown. Towards base of crown on labial side, ridges break up into distinctive dense reticulation. Teeth show considerable variation in shape along the jaw. Those from an anterior position are tall and upright with few lateral cusplets. Posteriorly, the crown profile lowers and the number of low lateral cusplets tends to increase.

Remarks. The teeth may be confused with those of '*Hybodus*' *minor*, being in the same size range and superficially similar in appearance. However, close study of their characteristics reveals many differences between the two species.

Distribution. *Synechodus rhaeticus* was originally described on the basis of a complete dorsal fin spine from the basal bone bed of the Westbury Formation at Aust Cliff, and a number of further fragments from the Holwell fissure fillings. Isolated teeth belonging to *Synechodus* have subsequently been discovered in bone bed deposits at a large number of sites both in Britain and continental Europe. They presumably belong to the same species as the dorsal fin spine fragments.

Genus NEMACANTHUS Agassiz, 1837
Nemacanthus monilifer Agassiz, 1837
Plate 27, figure 1

Description. Only isolated dorsal fin spines are known from this species. Length up to 100 mm, elongate, pointed at tip, cross section roughly triangular and laterally compressed. Anterior spine margin has rounded enamelled keel. Lateral walls of spine ornamented with small, round enamel tubercles which have patchy distribution and are absent from the spine apex, tending to cluster close to enamel keel about one-third from tip. Each posterolateral margin of spine bears longitudinal row of downturned denticles.

Remarks. *Nemacanthus* was first thought to be a ctenacanthiform shark (Maisey 1975), but Maisey (1977) placed it close to *Synechodus* within the Neoselachii.

Which teeth were associated with the fin spines of *Nemacanthus monilifer* is problematical. Woodward (1889a) suggested that '*Hybodus*' *minor* teeth belong to *Nemacanthus*, whilst Stensiö (1921, 1932) suggested that the teeth of *Polyacrodus* are better candidates.

Superorder GALEOMORPHII Compagno, 1973
Order LAMNIFORMES Berg, 1958
Family CETORHINIDAE Gill, 1862
Genus PSEUDOCETORHINUS Duffin, 1998a
Pseudocetorhinus pickfordi Duffin, 1998a
Plate 27, figure 2; Text-figure 20E–G

Description. Known only from isolated teeth up to 7 mm long. Cusp pointed, thorn-like, robust, variably inclined lingually and distally, situated centrally on crown. Lateral cusplets absent. Root projects lingually from crown underside, often flared at base. Labial face of root shallow. Tiny vascular foramina distributed randomly over root (anaulacorhize condition). Tooth morphology strongly variable through the dentition. Posterolateral teeth may develop cutting edges on central cusp and simple ornament of isolated, generally non-branching, short vertical striations. Distally through the dentition the central cusp is reduced and becomes increasingly inclined, incipient lateral cusplets develop and teeth become more elongate (Text-fig. 20E–G). Gill raker teeth (Pl. 27, fig. 2) up to 10 mm long, elongate, flattened, with slightly flared base and pointed, translucent tip.

Remarks. Preliminary investigation reveals that the enameloid ultrastructure consists of a single crystallite layer with a rudimentary parallel-fibred layer at the cutting edges, which suggests that *P. pickfordi* is a neoselachian shark. The range of dental morphology is closely similar to that described for *Cetorhinus* sp. from the Oligocene and Miocene of Belgium.

Assignation to the Cetorhinidae may require revision, but *Pseudo-cetorhinus* is undoubtedly the earliest recorded filter feeding shark. Extant cetorhinids are the only known passive filter feeding sharks, an average adult processing around 2000 tons of water hourly at cruising speeds of about 2 knots.

Distribution. A minor component of the Rhaetian ichthyofauna at the type locality of the species at Holwell Quarry, Somerset (Duffin 1998*a*).

CHONDRICHTHYAN SCALES

Remarks. The finer fractions of disaggregated bone bed and shale residues often yield large numbers of chondrichthyan scales. Shark scales are homologues of dentitional teeth representing a surprisingly complex organ system. Morphological changes take place as an individual ages and according to location on the body surface. They can be replaced when the skin is injured and their morphology seems to be related to shark ecology. Some morphotypes can be placed in higher taxonomic categories, but their diagnostic value to species level is difficult to assess at present. The taxonomic, ecological and potential biostratigraphical significance of chondricthyan scales remains largely unstudied.

Penarth Group chondrichthyan scales were first described by Sykes (1974*b*), with subsequent refinements made as part of a larger, more comprehensive study by Reif (1978).

Placoid scales
Plate 27, figure 3

Description. A single entrant vascular foramen perforates underside of base. Crown attached to base by means of a neck or pedicel which is often ridged. Crown spatulate, posteriorly directed. Central cusp usually present, may be flanked by up to two pairs of lateral cusps. Crown surface ornamented by number of keels which usually extend from cusp apices, often bifurcating as they pass onto scale neck. Base tetra-radiate to circular.

Remarks. These are non-growing scales and are the smallest of the chondrichthyan scales from the Penarth Group, with a maximum size of 3 mm. Placoid scales were found by Reif (1978) to be present in neoselachian sharks and pre-Rhaetian hybodont sharks, probably by convergence.

PLATE 27

Ctenacanthid scales
Plate 27, figure 4

Description. Base robust, undersurface concave. Short neck gives way to often multicuspid crown. Crown has between one and seven cusps, each posteriorly directed and ornamented by strong keels that branch close to crown/neck junction. Crown underside may be ornamented by fine striations. **Remarks.** These scales grew by lateral fusion and reached a maximum size of 5 mm, and are typical of ctenacanthiform sharks.

Hybodontoid scales
Plate 27, figures 5–6

Description. Base robust, relatively little expanded, surmounted by thick neck which supports unicuspid or multicuspid crown. Basal surface heavily rugose. Unicuspid specimens stud-like in appearance with very coarse keels ascending relatively short but very robust crown. **Remarks.** Multicuspid hybodontid scales are the product of odontode

---

EXPLANATION OF PLATE 27

Fig. 1. *Nemacanthus monilifer* Agassiz, 1837; fin spine; BMNH P 2854; 'Rhaetic Limestone, Somersetshire'; ×0·5.

Fig. 2. *Pseudocetorhinus pickfordi* Duffin, 1998a; gill raker tooth; BRLSI CD 1997/11; fissure infill=?Westbury Formation; Holwell, Somerset; ×5.

Fig. 3. Placoid scale; CJD collection; WB3 bone bed, Westbury Formation; Chilcompton, Somerset; anterior view; ×25.

Fig. 4. Ctenacanthoid compound scale; BRLSI CD 37; fissure infill=?Westbury Formation; Holwell, Somerset; anterior view; ×25.

Figs 5–6. Hybodont scales; fissure infill=?Westbury Formation; Holwell, Somerset. 5, single type; BRLSI CD 38; oblique lateral view; ×25. 6, compound type; BRLSI CD 39; lateral view; ×25.

Fig. 7. Hybodont cephalic spine; CJD collection; basal Westbury Formation bone bed; Aust Cliff, Gloucestershire; lateral view; ×1·5.

Figs 8–9. *Gyrolepis albertii* Agassiz, 1835; 8, tooth; AS 6219/24; Langport Member; Gotham Hills, near Nottingham; ×100. 9, scale; BMNH P 27646; 'Rhaetic bone bed'; Wainlode Cliff, near Gloucester; outer view; ×1·5.

Figs 10–12. *Severnichthys acuminatus* (Agassiz, 1835). 10, tooth of '*Birgeria acuminatus*' type; BMNH P 51430; Westbury Formation; Barnstone railway cutting, Nottinghamshire; lateral view; ×3. 11, tooth of '*Saurichthys longidens*' type; BMNH P 42461; 'Rhaetic Bone Bed'; Chipping Sodbury, Gloucestershire; lateral view; ×3. 12, jaw fragment with teeth; BMNH P 47287; Westbury Formation; Garden Cliff, Westbury-on-Severn, Gloucestershire; lateral view; ×0·75.

fusion in a temporal succession. Up to 12 individual crowns may be present on individual scales. Maximum size of individuals is 5 mm, but larger sizes are recorded in some Cretaceous forms, e.g. *Tribodus*.

Cephalic spines
Plate 27, figure 7

Remarks. Cephalic spines are paired structures located on the head of male hybodont sharks. Up to two pairs were present, the smallest located anteriorly above the supratemporal region of the braincase and the other posteriorly in the supraotic region. In older literature they are ascribed to the form genus *Sphenonchus*. They are believed to have been important for stimulation of the female during copulatory intertwining of the sexes.

The spines are asymmetrical and were anchored into the dermal tissues of the head by means of a triradiate base consisting of two lateral lobes and one medial lobe. The base is surmounted by a robust, posteriorly directed, recurved and enamelled crown. The base of the crown is ornamented by branching ridges.

Class HOLOCEPHALI Bonaparte, 1832

Remarks. Holocephalans (Text-fig. 19c), sometimes called chimaeras or ratfishes, are a group of cartilaginous fishes ranging from the Devonian to Recent. Like sharks, they possess a basal cartilage in the dorsal fin, scales with neck canals, and pelvic claspers in the male. A fin spine supports the first dorsal fin, whilst the long second dorsal fin is supported by radials only. The elongate whiplash-like tail has a long ventral fin behind the anus. Some groups possess an anal fin. The notochord is supported by a series of cartilage rings which develop within the notochordal sheath. There may be up to four notochordal sheath calcifications for each body segment. There is a complex pattern of sensory canals supported by tiny C-shaped cartilages. The most common fossils are isolated dorsal fin spines and the comparatively robust tooth plates. These are not replaced, but grow in the form of a logarithmic spiral. Most tooth plates have a crushing function, although some are adapted for cutting and slicing.

Order CHIMAERIFORMES Patterson, 1965
Suborder MYRIACANTHOIDEI Patterson, 1965
Family MYRIACANTHIDAE A. S. Woodward, 1889*a*

Remarks. Myriacanthids have dorsal fin spines with tuberculated walls and a row of downturned denticles along the postero-lateral spine margins. There are characteristically three toothplates in the lower dentition and up

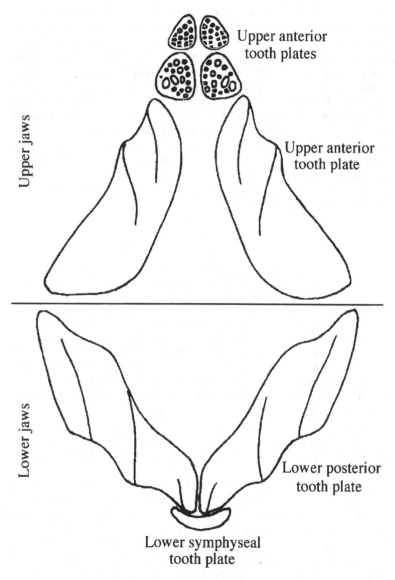

TEXT-FIG. 21. Myriacanthid holocephalan dental plates; reconstructed dentition looking into open mouth; lower posterior toothplate is *c.* 60 mm long.

to seven toothplates in the upper dentition (Text-fig. 21). A robust arched toothplate is located at the junction of the two lower jaws, succeeded by a pair of flat-surfaced posterior tooth plates, one on either side. Posterior upper toothplates are similar to the posterior lower toothplates but can be distinguished on the basis of their outline. There may be one or two pairs of triangular anterior upper toothplates. All possess a crown made up of a specialized tissue whose name has been the subject of some debate. It has been variously called orthotrabeculin, pleromin and pleromic hard tissue. It consists of a series of hypermineralized rods which are oriented normal to the crushing surface of the toothplate. This tissue generally has a broad distribution over the occlusal surface of the plate which may be divided by a series of diagonal ridges. The ridges lend a slicing component to the normally crushing toothplates. The base of each plate is usually well vascularized with blood vessels entering the plate from a shallow angle. The tooth plate was applied to the jaw by means of a thin, downwardly directed marginal ridge. Myriacanthoids are rare in the Penarth Group and are represented only by toothplates.

Genus AGKISTRACANTHUS Duffin and Furrer, 1981
Agkistracanthus mitgelensis Duffin and Furrer, 1981
Plate 28, figures 1–7; Text-figure 22

Description. Upper posterior toothplate (Pl. 28, figs 1–3) elongate sub-oval to trapezoid, reaching dimensions of 60 mm long by 20 mm wide. Symphyseal margin short. Labial margin accentuated by low ridge extending about 70 per cent. of length of toothplate. Deep wear facet clearly developed lingually due to biting against ridges on opposing posterior lower tooth plate. Hypermineralized tissue covers entire occlusal surface except for wear facet.

Lower toothplates (Pl. 28, figs 4–7) rare, possessing more rectangular outline with straight margins. Like upper toothplates, occlusal surface covered with hypermineralized tissue. Upper anterior toothplate unknown in this species.

Remarks. *Agkistracanthus* probably fed on hard-shelled bottom-living invertebrates, which it processed using its specialized crushing dentition. It may also have taken nektonic prey. Extant chimaeroids are often opportunistic feeders, living on the deeper parts of the continental shelf.

Distribution. *Agkistracanthus mitgelensis* was originally described from the Rhaetian and Hettangian of Switzerland and Austria on the basis of toothplates and dorsal fin spines (Duffin and Furrer 1981). It has been recorded in Britain from the Cotham Member at St Audries Bay and the Holwell fissure fillings, Somerset (Duffin 1994).

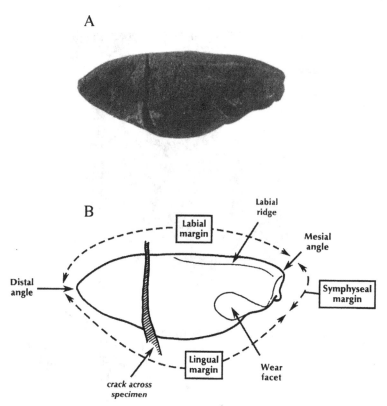

TEXT-FIG. 22. *Agkistracanthus mitgelensis* (Duffin and Furrer, 1981); right upper posterior toothplate; BMNH P 61526; occlusal view, Cotham Member; St Audrie's Bay, Somerset; specimen is *c*. 57 mm long (after Duffin 1994, fig. 5).

Genus MYRIACANTHUS Agassiz, 1837
Myriacanthus paradoxus Agassiz, 1837
Text-figure 23

Description. Lower toothplate trapezoid in occlusal view, measuring *c*. 40 mm diagonally. Like upper toothplate of *Agkistracanthus*, hyper-mineralized tissue covers entire occlusal surface. Most obvious diagnostic features are labial and lingual margins of plate and complex of diagonal ridges. In addition to long, thin labial ridge, *Myriacanthus* has robust, expanded central ridge and another ridge along the elongate symphyseal border.

A

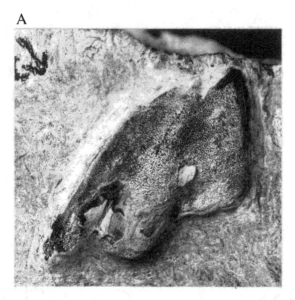

B

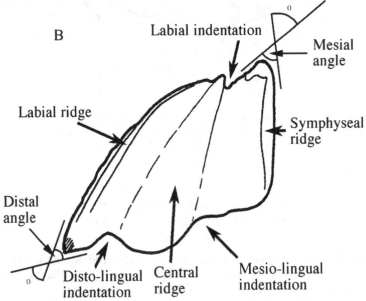

TEXT-FIG. 23. *Myriacanthus paradoxus* (Agassiz, 1837); left lower toothplate; BMNH P 62091; occlusal view, ?Upper Pecten Bed, Westbury Formation; Aust Cliff, Gloucestershire; longest diagonal is 37 mm across (after Duffin 1994, fig. 8).

Distribution. Only a single toothplate of this species has so far been described from the Penarth Group: a left lower plate believed to be from the 'Upper Pecten Bed', a prominent shell-rich limestone in the Westbury Formation of Aust Cliff (Duffin 1994). This species is better known, however, from isolated dorsal fin spines, toothplates and occasional articulated skulls from the Lower Jurassic of Lyme Regis (Patterson 1965).

Class OSTEICHTHYES Huxley, 1880

Remarks. Osteichthyes includes all fish with bony skeletons (Text-fig. 19D), and is the most diverse of all fish clades. Representatives of the class occur in both fresh and marine waters, ranging from Devonian to Recent. At their earliest appearance in the fossil record, they are already differentiated into two distinct lineages; the actinopterygians with fins supported by bony rays, and the sarcopterygians with fleshy lobes to their fins. In the Late Palaeozoic–Mesozoic, the actinopterygians diversified greatly, producing a wide range of body forms.

Subclass ACTINOPTERYGII Klein, 1885
Superdivision NEOPTERYGII Regan, 1925
Order PALAEONISCIFORMES Goodrich, 1909
Family PALAEONISCIDAE Vogt, 1852
Genus GYROLEPIS Agassiz, 1835
Gyrolepis albertii Agassiz, 1835
Plate 27, figures 8–9

Description. Teeth (Pl. 27, fig. 8) conical, up to 3 mm long. Base slightly flared. Apical cap enamelled, sharply pointed, unornamented, making up 35 per cent. of total tooth height. Shaft may possess superficial ornament of short, fine vertical striations.

Scales (Pl. 27, fig. 9) have rhomboid, glossy ganoin portion which would have been exposed in life. Exterior surface of ganoin bears branching series of fine ridges crossing from anterodorsal to posteroventral corner. Body of scale quadrilateral, with a projection forming an insertion prong at the anterior angle, providing anchorage in dermal tissues.

Remarks. *Gyrolepis* is represented in the Penarth Group by small teeth and isolated body scales. The conical teeth are common components of microvertebrate residues. Although defined by isolated scales and teeth, other species belonging to the genus are known from articulated material elsewhere e.g. the Middle Triassic of Germany (Dames 1888). For a recent account of the scale histology of *Gyrolepis albertii* see Richter and Smith (1995).

Distribution. *Gyrolepis albertii* occurs throughout the Penarth Group, although it is most common in the Westbury Formation. Also recorded from many Rhaetian localities in north-west Europe.

PLATE 28

Family *Incertae sedis*
Genus SEVERNICHTHYS Storrs, 1994
Severnichthys acuminatus (Agassiz, 1835)
Plate 27, figures 10–12

Description. Two types of teeth are associated with this species; each has previously been assigned to a separate taxon.

Teeth of '*Birgeria acuminatus*' type (Pl. 27, fig. 10) upright, conical, up to 15 mm long. A translucent enamel cap forms up to 50 per cent. of total tooth height. Cap separated from remainder of tooth by prominent ridge, may be rather compressed and bear lateral cutting edges. Fine vertical ridges ascend cap towards apex. Further ridges present beneath cap.

Teeth of '*Saurichthys longidens*' type (Pl. 27, fig. 11) upright but sigmoidal in outline. Translucent unornamented cap much smaller than '*Birgeria acuminatus*' type, often less than 10 per cent. of tooth length. Lower part strongly ridged. Base somewhat flared. Basal ridges may be strongly indented, giving folded structure to root wall.

Bone fragments assigned to *Severnichthys* have sculptured outer surfaces ornamented by many small tubercles. Rostropremaxillary bone very robust with double row of teeth. Small labial teeth flank much larger lingual tooth bases with a ratio of *c.* 3:1. Lingual tooth row consists of alternating tusk-like tooth bases with folded walls and intervening pits. Whole structure has roughly triangular outline with somewhat scalloped inner surface.

Lower jaw fragments (Pl. 27, fig. 12) have surface pattern of anastomosing tubercles and double row of teeth surmounting single broad shelf of bone. Row of large, robust lingual tusks and pits flanked labially by row of smaller teeth.

Remarks. This taxon is represented by isolated teeth and tooth-bearing bones. Isolated teeth are common components of the bone beds of the Westbury Formation, although larger bone fragments are known from some localities. *Severnichthys acuminatus* has a considerable synonymy which is briefly commented on below.

Severnichthys acuminatus was a large, predatory, bony fish which probably had a similar overall appearance to the well-known palaeoniscid genus *Birgeria*. Indeed, the conical teeth of the Penarth Group species

---

EXPLANATION OF PLATE 28

Figs 1–7. *Agkistracanthus mitgelensis* Duffin and Furrer, 1981; fissure infill=Westbury Formation?; Holwell, Somerset. 1–3, upper posterior toothplate; BRLSI M 186a; 1, occlusal view; 2, basal view; 3, lingual view. 4–7, lower posterior toothplates. 4, BRLSI M 186b; occlusal view. 5–6, BRLSI M 186e; 5, occlusal view; 6, basal view. 7, BRLSI M 186d; occusal view. All ×3·5.

were described by Stensiö (1921) as *Birgeria acuminata*. Fused rostropremaxillary and dentary bones bearing teeth have been found from the Aust and Westbury Garden cliff sections. These specimens have been very useful in clearing up some of the complex nomenclatural problems surrounding isolated teeth from the bone beds. Teeth described as *Birgeria acuminatus* and *Saurichthys longidens* occur together on the jaws, and there is a range of morphologies lying between the two extremes. In addition, folded walls of tusk-like tooth bases are found on the specimens. In the past some of this material, from a variety of localities, has been incorrectly identified as the teeth and jaws of labyrinthodont amphibians. Savage and Large (1966), and more recently Storrs (1994), have argued persuasively for the identity and chondrostean affinities of *Severnichthys* and consequently the absence of labyrinthodonts from the Penarth Group.

Distribution. Occurs throughout the Penarth Group, but as for almost all the fish remains, is commonest in the bone beds of the Westbury Formation.

Order PERLEIDIFORMES Berg, 1937
Family COLOBODONTIDAE Stensiö, 1916
Genus COLOBODUS Agassiz, 1844
Colobodus sp.
Plate 29, figure 1

Description. Both teeth and scales of *Colobodus* have been identified.

Teeth dome-shaped, of crushing type, up to 2 mm long and 2 mm in diameter. Circular to sub-oval in occlusal view, bearing central tubercle of acrodin (modified dentine), in centre of crown. Coarse vertical ridges ascend crown from point of junction with root. Root very short with wide pulp cavity at base. Bone to which teeth are attached quite deep, teeth closely spaced, often with eight to ten individual crowns of varying sizes clustered together.

• Scales (Pl. 29, fig. 1) usually fragmentary, rhomboid in outline, up to 10 mm in size. External face ornamented by series of coarse, sinusoidal, occasionally branching discontinuous ganoin ridges which cross external surface diagonally. Areas between ridges lack ganoin (Richter and Smith 1995), giving a distinctive morphology.

Remarks. The genus has a world-wide distribution but is confined to the marine Triassic.

Distribution. The British records are the youngest occurrence of the genus, which is otherwise a much more common faunal element in the European Muschelkalk. *Colobodus* sp. is an uncommon faunal component of the Penarth Group, only recorded with confidence from Aust Cliff, Holwell fissure fillings, and Spinney Hills and Glen Parva, both in Leicester (Browne 1890, 1895; Duffin 1980).

Division HALECOSTOMI Regan, 1923
Family SEMIONOTIDAE A. S. Woodward, 1890
Genus SARGODON Plieninger, 1847
Sargodon tomicus Plieninger, 1847
Plate 29, figures 2–3; Text-figure 19D

Description. Two types of teeth were carried by *Sargodon tomicus*.

Molariform teeth (Pl. 29, fig. 2) consist of isolated hemispherical crowns, circular to oval in occlusal view, measuring up to 7 mm in diameter, often heavily worn. Wearing reveals characteristic histology of underlying dentine, which consists of radiating network of large cavities with finely branching canaliculi at their ends. Molariform teeth arranged in longitudinal rows on both upper and lower jaws, probably with smallest teeth in front and at sides of dentitional pavement.

Incisiform teeth (Pl. 29, fig. 3) up to 15 mm long comprise chisel-like crown surmounting a deep root. Lingual face of crown divided into two by a wear facet in centre produced by functional ante-mortem contact abrasion. As in all bifid crowns, highest cusp located closest to midline of mouth.

Remarks. Other small (up to 2 mm long) hook-shaped teeth occur in microvertebrate residues and are possibly from dentigerous bones of the pterygoid complex of this species. *Sargodon* was originally erected on the basis of isolated teeth from the Rhaetian of Germany (Plieninger 1847), but has subsequently been recorded from Westbury Formation horizons throughout Britain. Well preserved, almost complete articulated specimens from the Norian of northern Italy (Tintori 1982) reveal *Sargodon* to have been a very deep-bodied fish measuring up to 1 m long which carried three to six incisiform teeth on the dentary and three teeth on the premaxilla. A reconstruction is given in Text-figure 19D. Scales have yet to be described from the Penarth Group, but the Italian material shows that they are deeper than broad, varying from rectangular to sub-pentagonal, with a smooth, unornamented outer surface. Incisiform teeth were probably used to pluck bivalves from the substrate, and the battery of molariform teeth provided an effective mill for breaking open their valves.

Distribution. Most common in the Westbury Formation, but occurs throughout the Penarth Group.

Genus LEPIDOTES Agassiz, 1832
Lepidotes sp.
Plate 29, figure 4

Description. A number of small teeth may be referable to the common Jurassic and Cretaceous genus *Lepidotes*. They are up to 1·5 mm in diameter and 1·7 mm long, dome-shaped, with oval to sub-circular outline.

PLATE 29

An off-centre tubercle made of acrodin commonly developed. Acrodin may be expanded to cover whole of crown surface, often preserved as isolated tooth caps in bone bed deposits. Surface of crown not striated or otherwise ornamented.

Remarks. Small, hook-like, branchial teeth found in bone bed microvertebrate faunas may also belong to this genus. *Lepidotes* is a well-known genus ranging from the Rhaetian to the Upper Cretaceous of Europe, the Americas and India. Records from the Penarth Group are the oldest (Jain 1983). The large glossy rhomboidal scales of *Lepidotes* so characteristic of Jurassic occurrences of the genus are unknown in the Penarth Group, suggesting that the Penarth Group material may possibly be referable to a fish other than *Lepidotes*.

Distribution. Paired vomers attributed to *Lepidotes* sp. have been recorded from the Penarth Group of Watchet by Jain (1983), but the genus has otherwise been described only from isolated teeth recorded from Barnstone, Nottinghamshire (Sykes 1979). Unsubstantiated records were documented by Moore (1861*b*) and Richardson (1911).

Family MACROSEMIIDAE Thiolliére, 1858
Genus LEGNONOTUS Egerton, 1854
Legnonotus cothamensis Egerton, 1854
Plate 29, figure 5

Description. Originally described from complete, articulated specimens, this fish is small, up to 60 mm long, shallow-bodied, covered with tiny

---

rhomboid scales with coarsely serrated posterior margins. These are absent from dorsal fin region. Large dorsal fin approximately half length of body, supported by about 30 fin rays. Tail homocercal, i.e. symmetrical about the long axis, with relatively broad junction with body. Dentition little known, comprising tiny blunt teeth.

Remarks. Complete fish fossils are extremely rare in the Penarth Group; this is one of only two known taxa, the other being *Pholidophorus higginsi*. Bartram (1977) placed this genus into the macrosemiid holosteans on the basis of dentary structure. *Legnonotus* is the earliest recorded member of the group, which ranges into the Upper Cretaceous.

Distribution. *Legnonotus cothamensis* is known on the basis of a few specimens from the Cotham Member of the Lilstock Formation at Aust Cliff (Egerton 1854).

Subclass TELEOSTEI Müller, 1846*b*
Family PHOLIDOPHORIDAE A. S. Woodward, 1891
Genus PHOLIDOPHORUS A. S. Woodward, 1891
Pholidophorus higginsi Egerton, 1854
Plate 29, figure 6

Description. Body fusiform, maximum length 65 mm, four times depth in trunk region. Rhombic scales smooth, thick and deeply imbricating, posterior margins somewhat rounded, occasionally produced into four slender, pointed, posteriorly directed denticles. Flank scales produced in four longitudinal series, much deeper than broad. Opercular apparatus on head occupies one-quarter of length of the fish. Skull bones enamelled. Pelvic fins located in middle of the trunk, approximately same distance below as dorsal fin.

Remarks. *Pholidophorus higginsi* is one of the oldest representatives of the genus, which ranges from the Upper Triassic to the Upper Jurassic (Gardiner 1960). In section, the scales possess a thick outer enamel (ganoin) layer overlying an even thicker layer of lamellar bone (Gardiner 1960). The proportions of this Late Triassic species are much the same as those of extant herring or smelt but, with enamelled skull bones, it belongs to a completely different teleost group.

Distribution. Like *Legnonotus cothamensis*, this fish was first described from the Cotham Member of the Lilstock Formation at Aust Cliff. Egerton (1854) described a second species from Aust as *Pholidophorus nitidus*, but this was subsequently synonymized with *P. higginsi* (Woodward 1889*b*). The species has also been recorded from Wigston (Glen Parva), Leicestershire (Harrison 1876; Woodward 1889*b*, 1895).

Class SARCOPTERYGII Romer, 1955
Infraclass DIPNOIFORMES Cloutier, 1993

Order DIPNOI Müller, 1844
Suborder CERATODONTOIDEI Vorobyeva and Obruchev, 1964
Family CERATODONTIDAE Gill, 1872
Genus CERATODUS Agassiz, 1838
Ceratodus latissimus Agassiz, 1839
Plate 29, figure 7; Text-figure 5

Description. Most fossil material of *Ceratodus* comprises very variable, isolated, robust toothplates, occasionally still attached to bone. Plates crescentic, measuring up to 100 mm long. Labial margin is longest edge. Strong ridges transect plate from mesial margin to rather crenulated labial margin, giving series of troughs and ridges over occlusal surface. Surface punctuated by closely spaced pores.

Remarks. Ten species of *Ceratodus* were originally erected for the material described by Agassiz (1839) from the Aust bone bed, but these have since been reduced to a maximum of two: *C. latissimus* and *C. parvus*. Toothplates of *C. parvus* are smaller than those of *C. latissimus*, and it is unclear if *C. parvus* is a junior synonym based on juvenile variants, or a species in its own right (Storrs 1994). Lungfish toothplates are distinctive elements of the bone bed faunas; indeed, in older literature, the bone beds of Aust and Westbury-on-Severn where *Ceratodus* is most common, although never abundant, were occasionally called the 'Ceratodus Bed'.

Distribution. Westbury Formation bone beds in Gloucestershire and Leicestershire (Woodward 1891). A recent discovery at Blue Anchor Bay, Somerset may comprise articulated lungfish remains (Pl. 25), but a Penarth Group horizon has yet to be confirmed.

Infraclass ACTINISTIA Cope, 1871
Suborder COELACANTHOIDEI Berg, 1937
Family COELACANTHIDAE Agassiz, 1843
Undetermined coelacanth
Plate 29, figure 8

Description. Coelacanthids are known only on the basis of isolated quadrate bones from the skull. These are distinctive structures comprising a 15 mm high, rod-like element with double articulation at base. Central, elongate portion may be produced into a broad flange in more complete specimens.

Remarks. The quadrate bones described here provide the first, rather tantalising, evidence of coelacanthids from the Penarth Group.

Distribution. The only record so far of coelacanthid quadrates from the Penarth Group is a specimen figured (and incorrently identified as a reptilian phalange – a common error) from Holwell (Duffin 1978*a*, fig. 7).

DERIVED FISH FOSSILS

Fossils derived from underlying Palaeozoic rocks and incorporated into Penarth Group sediments have been recorded from a few localities. Various bradyodont holocephalan toothplates have been recorded from the main Westbury Formation bone bed at Aust, including *Cochliodus*, *Psammodus*, *Psephodus* and *Helodus* (Storrs 1994). A record of *Ctenoptychius* (Davis 1881) from the same locality is a misidentification of a neural spine of the choristodere *Pachystropheus rhaeticus*.

Duffin (1993c) reported a diverse fish fauna derived from the Carboniferous and incorporated in sediments of Rhaeto-Liassic age which fill palaeokarst fissures at Tytherington Quarry, Gloucestershire. Preliminary identification indicates the presence of acanthodian scales, osteichthyans and a wide range of chondrichthyans (phoebodontids, stethacanthids, protacrodontids, orodontids, psammodontids and early neoselachians), including a number of new taxa.

NOTE ADDED IN PRESS

Two important papers have recently appeared concerning the enameloid structure of sharks' teeth from the Penarth Group. Cuny (1998) and Cuny and Benton (1999) described triple-layered enameloid in '*Hybodus*' *minor*, indicating that it is a neoselachian. *Pseudocetorhinus pickfordi* is also shown to have triple-layered enameloid, but with a very thick single crystallite layer, confirming its allocation to the Neoselachia (Cuny and Benton 1999). *Vallisia coppi*, however, has only a single crystallite enameloid layer, which, together with a distinctive morphology, suggests that it is neither a neoselachian nor a hybodont shark (Cuny and Benton 1999).

15. TETRAPODS

by GLENN W. STORRS

Tetrapod remains are abundant in the bone beds of the Westbury Formation of the Penarth Group in south-west Britain, yet form a mixture of familiar and enigmatic fossils. Partly, this is a function of the taxonomic composition of the fauna, containing as it does several problematical taxa. It is also a reflection of the mixed terrestrial and marine faunas present in these bone beds and the taphonomic factors involved in their preservation. Perhaps the most influential factor, however, is the universally disarticulated nature of the vertebrate remains and the extreme rarity of any suggestion of associated individuals. The disarticulation, mixing, difficulty in preparing bone bed samples and the taxonomic complexity of the tetrapod fauna combine to create a poorly known and inadequately studied suite of fossils.

Past reviews of the vertebrates from the Penarth Group, and in particular those from the Westbury Formation, include those of Reynolds (1946), Sykes *et al.* (1970), Sykes (1977), Duffin (1978*b*) and Storrs (1993, 1994). The majority of the tetrapod fossils found in the bone beds comprises teeth, vertebral centra and limb bones. Partial ribs are also common, but limb girdle and skull bones are rare. Various taphonomic biases are apparent in the distribution of remains, as noted by Storrs (1994) and Storrs *et al.* (1996).

Among the most readily recognizable elements are isolated vertebrae and other fragments of ichthyosaurs and plesiosaurs. Because of the generally conservative nature of the postcranial elements of these marine reptiles, however, it has often proved difficult to identify accurately the genera represented by such bones. The other major tetrapod components common in the bone beds of the Westbury Formation in particular, but also occurring above and below this unit, are vertebrae and propodials (humeri and femora) of the early choristodere, *Pachystropheus rhaeticus*. Ichthyosaur, plesiosaur and choristoderan remains occur commonly at Aust Cliff, Garden Cliff (Westbury-on-Severn) and Newark, Nottinghamshire (Martill and Dawn 1986; Storrs 1994; Storrs *et al.* 1996). *Pachystropheus* is particularly abundant at Garden Cliff. Other tetrapod remains are uncommon to very rare throughout the Penarth Group.

Tetrapod remains potentially of Rhaetian age are known from fissure fillings of the Mendip archipelago and southern Wales. These important faunas are notoriously difficult to date because of their isolated nature and estimates range from Carnian to Sinemurian (Benton 1984). Some of the fossils collected from these fissure systems include taxa that are represented in the Penarth Group, although the possibility of reworking cannot be ignored. The most likely candidates for fissure deposits of an equivalent age to those of the Penarth Group, i.e. Rhaetian, are at Holwell, near Frome, Somerset, and Tytherington and Cromhall (formerly Slickstones) in south Gloucestershire (Storrs 1994). The palaeontology of these deposits has been well reviewed by Kühne (1946), Robinson (1957), Savage and Waldman (1966), Fraser (1985, 1986), Whiteside (1986) and Storrs (1994).

Specimens on Plate 30 are held in The Natural History Museum, London (figs 1, 3, 6), Bath Royal Literary and Scientific Institute (figs 2, 5) and Bristol City Museum and Art Gallery (fig. 4)

MARINE REPTILES

The vast majority of tetrapod remains recovered from the bone beds of the Penarth Group are those of secondarily aquatic ones, specifically marine reptiles. Several species have been described from such remains, but few are determinable on the basis of the known specimens. The relatively common ichthyosaurs and plesiosaurs are representative of two independent marine lineages. Each group had terrestrial forebears and seems to be derived ultimately from diapsid reptiles, although the pleisosaurs had become so specialized that the cheek bar closing the lower temporal opening of the skull has been lost (Kuhn-Schnyder 1980).

Class AMNIOTA Haeckel, 1866
Subclass DIAPSIDA Osborn, 1903
Order ICHTHYOSAURIA de Blainville, 1835

Remarks. Ichthyosaurs (Text-fig. 24A1), or 'fish reptiles', are a classic example of convergent evolution, having adopted a fusiform bauplan that parallels those of fish and cetaceans. Virtually all ichthyosaur fossils from the Penarth Group are generically indeterminate, being largely isolated disc-like and deeply amphicoelous vertebral centra with double rib facets (Text-fig. 25A–B), ribs, paddle elements (flattened blocks and discs with unfinished edges) and conical teeth with expanded and coarsely ribbed roots (Text-fig. 24A2). Oddly, the triangular lachrymal bone forming the anterior lower corner of the eye socket is relatively commonly encountered in the Westbury Formation as an isolated element. This is perhaps a factor of its own peculiar hydrodynamic characteristics as it was transported and perhaps winnowed by wave or current action from larger bone accumulations. The very large size of some ichthyosaur bones in the Westbury Formation is notable.

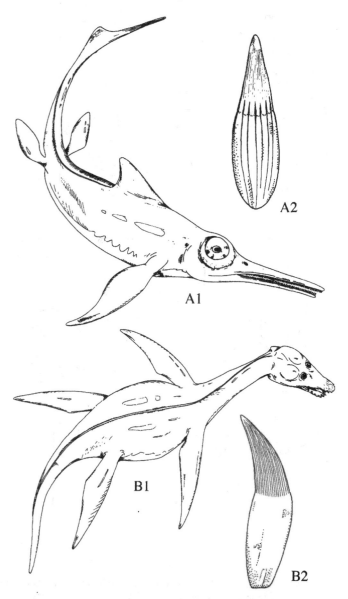

TEXT-FIG. 24. A1, *Leptonectes* (McGowan, 1996); life reconstruction of Rhaetian–Early Jurassic ichthyosaur. A2, generic ichthyosaur tooth. B1, *Eurycleidus* Andrews, 1922; life reconstruction of Rhaetian–Early Jurassic plesiosaur. B2, generic plesiosaur tooth. Not to scale.

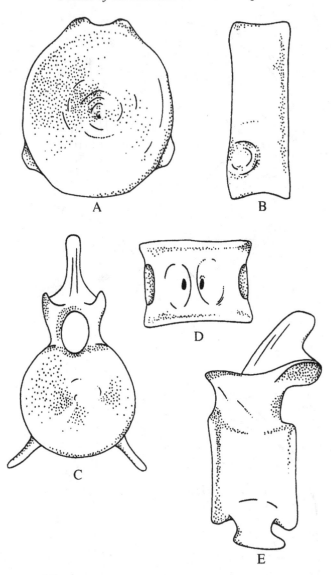

TEXT-FIG. 25. Morphology of isolated marine reptile vertebrae. A, anterior caudal centrum of indeterminate ichthyosaur in anterior view; note loss of neural arch. B, same in lateral view. C, cervical vertebra of indeterminate plesiosaur in anterior view; arch is often fused to centrum. D, same in ventral view; cervical ribs lacking (these may be fused or unfused); note characteristic nutrient vessel openings. E, same in lateral view. Not to scale.

Family LEPTOPTERYGIIDAE Kuhn, 1934
Genus LEPTONECTES McGowan, 1996
Leptonectes sp.

Description. 'Latipinnate' ichthyosaur of moderate size, reaching lengths of 2·5 m or more. Long slender snout with slight overbite. Vertebral column with only slight or indistinct downward bend of tail. Teeth often

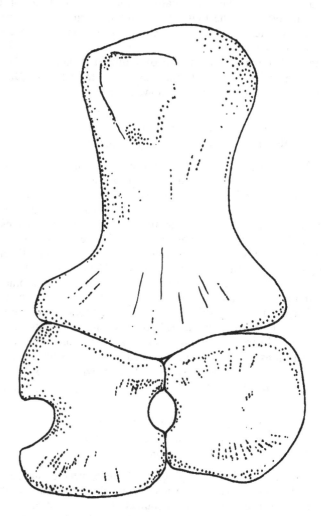

TEXT-FIG. 26. *Leptonectes tenuirostris* (Conybeare, 1822); proximal forearm: humerus, radius and ulna; ×0·5 (after McGowan 1989, text-fig. 2c).

slender and small. Perhaps only four digits to forelimb. Humerus slender at mid-shaft, but broadly expanded distally. Radius notched.

Remarks. Only a single humerus from the Penarth Group has to date been identified as belonging to *Leptonectes*. This is from the Westbury Formation of Chipping Sodbury, Gloucestershire (Storrs 1994). It is the only generically determinate ichthyosaur fossil so far identified from the Penarth Group. A reconstruction of a generic type of *Leptonectes* is given in Text-figure 24A1. Ichthyosaur remains become more common in the basal Rhaetian portion of the Lias Group, and a drawing of a proximal forearm of the species *L. tenuirostris* (Conybeare) from these levels is illustrated in Text-figure 26.

Distribution. Rhaetian to Sinemurian of southern Britain: principally Lower Lias of Street in Somerset, but also known from Devon, Dorset, Gloucestershire and Leicestershire (McGowan 1989).

Order SAUROPTERYGIA Owen, 1860
Suborder PLESIOSAURIA de Blainville, 1835

Remarks. Plesiosaur remains have previously been considered to be the most common tetrapod fossil found at the classic locality of Aust Cliff (Wickes 1904). In general, however, they are found far less frequently than ichthyosaur bones. No plesiosaur from the Penarth Group can be generically, even less specifically, identified, despite several species of '*Plesiosaurus*' having been named at least partly on the basis of Westbury Formation material (Owen 1840; Swinton 1948; Storrs 1994). Each of these species is now considered to be a *nomen dubium* (Storrs 1994). The vertebrae are roughly cylindrical with flat ends and paired ventral foramina and the neural arches are almost always fused (Text-fig. 25c–e). The teeth are large, sharp, curved and striated (Text-fig. 24b2) and along with the vertebrae constitute the most numerous remains. Limb elements, such as partial humeri and spool-shaped phalanges, are also commonly found. Supposed gastroliths occur as small, polished quartzose pebbles (Neville 1967), but no direct evidence exists to link them with plesiosaurs, nor indeed with any other organism.

Superfamily PLIOSAUROIDEA Nopcsa, 1928
Family RHOMALEOSAURIDAE Kuhn, 1961?
Gen. et sp. indet.

Remarks. Although no plesiosaur genus can be recognized on the basis of the currently known Penarth Group fossils, the short cervical vertebrae and the large, robust teeth typically encountered suggest that big-headed, relatively short-necked forms similar to the British Rhaetian–Lower Jurassic genus *Eurycleidus* Andrews, 1922 may have been their source. A life reconstruction of a generic type of *Eurycleidus* is given in Text-figure 24B1.

Cruickshank (1996) noted a primitive plesiosaur humerus from the Rhaetian–Hettangian of Barrow upon Soar, Leicestershire, whose large size suggests that it is the propodial of a pliosauroid. It too is generically indeterminate.

Suborder PLACODONTIA Owen, 1859?
Family PLACOCHELYIDAE Jaekel, 1907?
Genus PSEPHODERMA von Meyer, 1867?
Psephoderma anglicum von Meyer, 1867?

Remarks. In rare cases, the Westbury Formation has yielded scute-like bones resembling those from the Holwell fissure deposits described by von Meyer (1867) as *Psephoderma anglicum*, in the belief that these were placodont remains akin to the Upper Triassic *P. alpinum* of the Alpine province. Placodonts were bottom-foraging sauropterygian reptiles related to plesiosaurs, but characterized by a short neck, crushing dentition and in the case of *Psephoderma* and some others, a dorsal carapace of interlocking scutes. The English material is problematical in that no obvious placodont bones are known from British rocks, although teeth attributed to placodonts have been reported from Holwell (Moore 1859, 1861a; Kühne 1946; Duffin 1978b).

Order CHORISTODERA Cope, 1876
Family PACHYSTROPHEIDAE Kuhn, 1961
Genus PACHYSTROPHEUS E. von Huene, 1935
Pachystropheus rhaeticus E. von Huene, 1935
Plate 30, figures 1–5

Description. Primitive crocodile-like choristodere with relatively elongate vertebrae with flattened ends, broad neural arch facets and generally rectangular neural spines with thickened, wrinkled tips. Neural arches/spines typically separated from centra due to lack of co-ossification. Humeri flattened with sharp postaxial ridge along their length. Femora rather straight with rounded proximal head and broad distal end.

Remarks. Storrs *et al.* (1996) described the significant specimens of *Pachystropheus rhaeticus* known to date. *Pachystropheus rhaeticus* was described by E. von Huene (1935) from specimens from Rhaetian (Westbury Formation) bone beds of south-west England (Somerset and Gloucestershire), Germany and from the Holwell fissure fillings. It is probable that *Pachystropheus* is synonymous with *Rysosteus oweni* Woodward and Sherborn 1890, a name commonly found in the local literature. The genus was coined by Owen (1842) on the basis of a single

PLATE 30

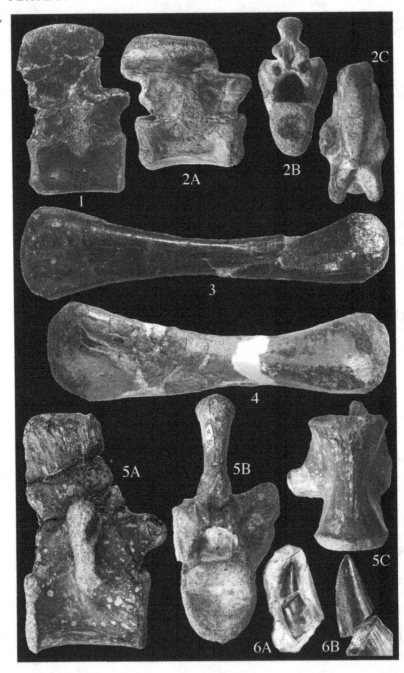

vertebra from Aust Cliff with supplementary material from Aust and Garden Cliff (Duffin 1978*b*). However, the short descriptions, lack of illustration and loss of the holotype make *Rysosteus* a *nomen dubium* (Storrs and Gower 1993; Storrs *et al.* 1996).

Long an enigmatic species, *P. rhaeticus* is now thought to be an early choristodere, thus extending back in time the stratigraphical range of this group by almost 40 million years (Storrs and Gower 1993; Storrs 1994; Storrs *et al.* 1996). Specimens attributable to *Pachystropheus* are relatively numerous in bone bed deposits, especially those at Aust and Garden cliffs. Features suggestive of choristoderan affinities are the thickened ribs and gastralia (belly ribs), flat-ended vertebrae, deep yet constricted neural canal, double-headed dorsal ribs articulating to both the neural arch and vertebral centrum, L-shaped clavicle, interclavicle with stout posterior process, coarsely ribbed ilium, flattened humerus with distinct ectepicondylar groove on outer edge of distal end and straight femur with prominent internal trochanter and indistinct articular surfaces, among others.

While most specimens of *Pachstropheus* suggest an animal about 1 m long or less, rare examples indicate adult lengths of 2–2·5 m. Except for *Pachystropheus*, all other known choristoderes are from freshwater, usually alluvial, deposits. *Pachystropheus* specimens from marginal marine rocks may represent coastal or estuarine animals.

Distribution. Westbury Formation of England and Wales, but also Rhaetian of Germany and Rhaetian or Norian of France. One humerus has been recovered from the lower beds of the Lias Group (Hettangian) in Somerset (Storrs *et al.* 1996).

TERRESTRIAL REPTILES

Whilst marine animals such as sharks, plesiosaurs and ichthyosaurs dominate the Penarth vertebrate fauna, occasional terrestrial components

---

EXPLANATION OF PLATE 30

Figs 1–5. *Pachystropheus rhaeticus* E. von Huene, 1935; postcranial bones; Westbury Formation. 1, BMNH R 6851; average dorsal vertebra in lateral view; Garden Cliff, Westbury-on-Severn, Gloucestershire. 2a–c, BRLSI C 28; anterior dorsal vertebra in lateral, posterior and dorsal views; Vallis Vale, near Frome, Somerset. 3, BMNH R 747; left femur in dorsal view; Blue Anchor Point, Somerset. 4, BCMAG Cb 4907; left humerus in ventral view; main basal bone bed; Aust Cliff, Gloucestershire. 5a–c, BRLSI M 205; large dorsal vertebra in lateral, posterior and ventral views; Vallis Vale, near Frome, Somerset. All ×0·75.

Fig. 6. '*Picrodon herveyi*' tooth; BMNH R 2875; 'Rhaetic'; Wedmore Hill, Somerset; 6a ×1; 6b ×2·5.

such as dinosaurs (cf. *Camelotia, Megalosaurus*?) and freshwater lungfish (*Ceratodus*) are also found. They suggest a nearby fluvial contribution to the deposits (Storrs 1993, 1994). The nearshore palaeoenvironment of the Westbury Formation meant that on occasion, the bodies or bones of land-living reptiles, particularly dinosaurs, were washed out to sea by flood events or natural river action. These were subsequently buried, reworked by storms, frequently abraded and redeposited.

Subdivision ARCHOSAURIA Cope, 1869
Superorder DINOSAURIA Owen, 1842

Remarks. Dinosaur bones are occasionally found in the Westbury Formation but, by virtue of their usually fragmentary nature, generally cannot be identified more precisely than Dinosauria indet. Many are recognized only by their anomalously large size, as in the case of hindlimb bones, for example. Only rarely does a smaller bone retain sufficient anatomical detail for identification as dinosaurian.

Order SAURISCHIA Seeley, 1887
Suborder SAUROPODOMORPHA F. von Huene, 1932
Family MELANOROSAURIDAE F. von Huene, 1928
Genus CAMELOTIA Galton, 1985
Camelotia borealis Galton, 1985
Text-figure 27A–D

Description. A large prosauropod dinosaur (*c.* 9 m long) with pleurocoels (excavated, possibly pneumaticized, cavities) in its larger dorsal vertebral centra. Its femoral shaft is straight and robust.

Remarks. Dinosaur bone fragments in the Westbury Formation have often been attributed to *Avalonia* (e.g. Macfadyen 1970; Hamilton 1977). *Avalonia*, however, is a 'waste-basket taxon' and furthermore, a pre-occupied name. Kuhn (1961) proposed *Avalonianus* as a replacement. *Avalonianus* (*Avalonia*) *sanfordi* was described from a serrated tooth from the Wedmore Stone facies of the Westbury Formation, Wedmore Hill, Somerset (Seeley 1898). Seeley (1898) also assigned a partial skeleton from the same locality to this species, whereas additional vertebrae were assigned to another new genus and species, *Picrodon herveyi*, founded upon a second tooth (Pl. 30, fig. 6). Actually, all of the postcranial bones are merely different parts of the same animal (Newton 1899). These are now known (Galton 1985) as *Camelotia borealis*. This is one of the very rare associations of bones of a single individual and the only dinosaurian association from the Westbury Formation.

   Camelotia is one of the largest prosauropods known and is the only melanorosaurid recorded from the northern hemisphere (Galton 1985).

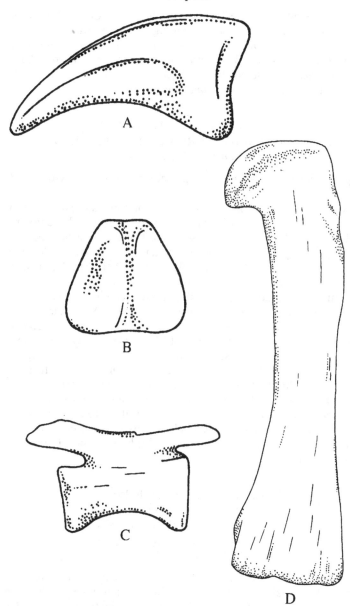

TEXT-FIG. 27. *Camelotia borealis* Galton, 1985; postcranial bones. A–B, ungual phalanx or claw of foot in lateral and proximal aspects; ×0·25. C, caudal vertebra; ×0·25. D, left femur; ×0·05 (after Seeley 1898, pl. 1, figs 4, 5, 3, 1, respectively).

The carnivore teeth associated with the holotype of *Camelotia* are the probable result of predation or scavenging and thus *Avalonianus* and *Picrodon* are *nomina dubia* (Van Heerden 1979). Much of the large dinosaur material from the Westbury Formation (primarily from Aust), although generically unidentifiable, such as the bone (femur?) reported by Stutchbury (1850), may be from *Camelotia*.

Family ANCHISAURIDAE Marsh, 1885?
Genus THECODONTOSAURUS Riley and Stutchbury, 1836
Thecodontosaurus antiquus Morris, 1843

Description. Small (*c*. 2–3 m long) prosauropod dinosaur with gracile limbs.

Remarks. *Thecodontosaurus antiquus*, a second prosauropod from the Upper Triassic of the Bristol Channel region and a presumed anchisaurid, is based upon a partial jaw from the so-called Dolomitic (or 'Magnesian') Conglomerate at Clifton, Bristol (Riley and Stutchbury 1836). This deposit is actually a fissure or cave collapse deposit (Halstead and Nicoll 1971). Small prosauropod remains are occasionally found in Carboniferous Limestone fissures in Somerset, Gloucestershire and Glamorgan and usually are called *Thecodontosaurus* (e.g. Moore 1859; Marshall and Whiteside 1980; Benton 1984; Kermack 1984). However, the age(s) of these deposits is problematical and whilst Moore (1881) regarded them as 'Rhaetic', they are often considered to be ?Norian–Rhaetian in age. A discussion would not be relevant here, were it not for the fact that Weishampel (1990) lists *Thecodontosaurus* sp. as occurring in the Westbury Formation of Somerset. It has not been established unequivocally that any material from the Penarth Group belongs to *Thecodontosaurus*, but many of the smaller dinosaur bones could possibly be attributed to this genus. Benton *et al.* (1999) have thoroughly reviewed *Thecodontosaurus*.

Suborder THEROPODA Marsh, 1881
Family MEGALOSAURIDAE Huxley, 1869?
Genus MEGALOSAURUS Buckland, 1824?
Megalosaurus? *cambrensis* (Newton, 1899)
Text-figure 28

Description. Mid-sized (*c*. 5 m long), bipedal, carnivorous dinosaur.
Remarks. A natural 160 mm long mould of a lower jaw with sharp, serrated teeth from the 'Rhaetic' of Glamorgan, South Wales was described by Newton (1899) as *Zanclodon cambrensis*, although *Zanclodon* is an undiagnosable genus. The type specimen of *Zanclodon* from Germany is an indeterminate tooth (Benton 1986). *Z. cambrensis* has subsequently

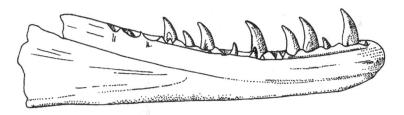

TEXT-FIG. 28. *Megalosaurus*? *cambrensis* (Newton, 1899); lower jaw; ×0·1 (after Newton 1899, pl. 10, fig. 1). Curved, laterally compressed teeth with serrated edges are typical of carnivorous dinosaurs.

been referred to as '?*Megalosaurus cambrensis*' by Weishampel (1990). The age of the sandstone block containing the fossil is uncertain, but it is probably from a local near-shore facies on Stormy Down, possibly equivalent to the Lilstock Formation elsewhere (Ivimey-Cook 1974). *Megalosaurus*? *cambrensis* is an early theropod whose precise systematics are obscure. Occasional theropod vertebrae from Aust and Lavernock may belong to this species.

Order PARASUCHIA Huxley, 1875?
Family PHYTOSAURIDAE Lydekker, 1888?
Genus PALAEOSAURUS Riley and Stutchbury, 1836? *non* Geoffroy, 1833
'?*Palaeosaurus stricklandi*' Davis, 1881

Remarks. Davis (1881) referred a non-diagnostic tooth from the 'Rhaetic bone bed' of Coombe Hill, near Cheltenham, Gloucestershire, to the new species '?*Palaeosaurus stricklandi*', now a *nomen dubium*. Riley and Stutchbury (1836) had earlier named two serrated teeth from the *Thecodontosaurus* type locality *Palaeosaurus cylindrodon* and *P. platyodon*. These also are essentially indeterminate. All represent archosaurian predators of questionable relationship (Charig *et al.* 1965). F. von Huene (1908) considered *P. platyodon* to be phytosaurian and possibly allied to his *Rileya bristolensis* (*Rileya* is preoccupied and replaced by *Rileyasuchus* Kuhn, 1961). The latter is possibly a phytosaur (crocodile-like archosaur from the Triassic) based upon postcrania from the Clifton *Thecodontosaurus* deposit (F. von Huene 1902). Benton *et al.* (1999) refer most of the *Rileyosuchus* material to *Thecodontosaurus*.

Order *Incertae sedis*
Genus SAURODESMUS Seeley, 1891

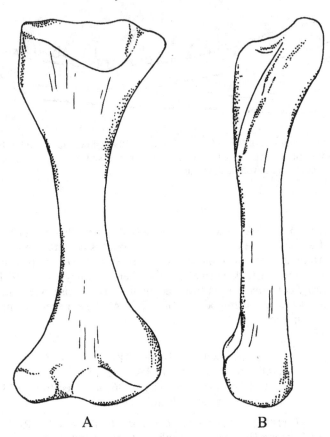

TEXT-FIG. 29. *Saurodesmus robertsoni* Seeley, 1891; right humerus; A, dorsal view.
B, postaxial view. Both ×0·5 (after Seeley 1891, figs 1, 6).

Saurodesmus robertsoni Seeley, 1891
Text-figure 29

Remarks. A vertebrate assemblage resembling that of the Westbury
Formation is known from Linksfield, near Elgin, Morayshire, Scotland.
Historically considered 'Rhaetic', it belongs to a limestone and marl
sedimentary outlier, considered to be a very large glacial erratic (Taylor
and Cruickshank 1993). Presumably originating offshore to the north of
the Scottish coast, the unit contains darwinulid ostracods which are
possibly of Rhaetian age (Anderson 1964). Among the Linksfield
vertebrate fauna, a single problematical humerus named *Saurodesmus
robertsoni* represents a reptile of unknown affinity. Seeley (1891)

identified it as a crocodylian and Duff (1842) as a turtle. Neither is correct and it must represent an unknown terrestrial or littoral component of the latest Triassic fauna. It bears some resemblance to rhynchosaurs, but is apparently too large and would represent a significant range extension of this group.

MAMMALS

The rarest of all tetrapod remains in the Penarth Group are those of early mammals. Apart from a few dozen specimens from questionably Rhaetian fissure fill deposits, only two apparent mammals and one mammal-like therapsid are known from the stratified deposits of the Penarth Group.

Subclass SYNAPSIDA Osborn, 1903
Order THERAPSIDA Broom, 1905
Suborder CYNODONTIA Owen, 1860?
Family *Incertae sedis*
Genus TRICUSPES E. von Huene, 1933?
Tricuspes? sp.

Remarks. A single small tooth was reported as *Tricuspes* from the Westbury Formation at Vallis Vale, Somerset, by E. von Huene (1933). *Tricuspes*, from the German Rhaetian, may be a mammal with affinities to the Triconodonta (Clemens *et al.* 1979; Clemens 1980) or perhaps part of the pre-morganucodontid radiation (Storrs 1994). E. von Huene (1933) considered *Tricuspes* to be cynodontian. There is some doubt that the English specimen is truly *Tricuspes*, but may rather be a distinct therapsid.

Order ALLOTHERIA Marsh, 1880?
Family HARAMIYIDAE Simpson, 1947
Genus HARAMIYA Simpson, 1947
Haramiya antiquus (Plieninger, 1847)
Haramiya fissurae (Simpson, 1928)
Text-figure 30

Description. A small, shrew-like mammal with low-crowned, multicuspate teeth.
Remarks. Teeth of *Haramiya* are known primarily from the Holwell fissure fills from which two species have been described [*H. fissurae* (Simpson, 1928) and *H. antiquus* (Plieninger, 1847)]. Etheridge (1872) figured a tooth crown, apparently of *H.* (*Microlestes*) *antiquus*, from the Westbury Formation near Penarth. Haramyids are currently considered to

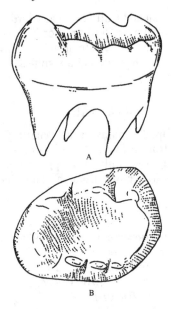

TEXT-FIG. 30. *Haramiya fissurae* (Simpson, 1928); reconstructed molariform tooth; A, lateral view, B, occlusal view; Rhaetian fissure fillings; Holwell, Somerset; ×7·5 (after figures in Simpson 1928).

be basal mammalian-grade animals, with a possible link to multi-tuberculates (Clemens and Kielan-Jaworowska 1979).

Family HARAMYIDAE Simpson 1947?
Genus HYPSIPRIMNOPSIS Dawkins, 1864
Hypsiprimnopsis rhaeticus Dawkins, 1864

Remarks. This taxon was described (Dawkins 1864, pp. 409–412, figs 3–5) from a single tooth from the 'Sully Beds', a grey marl facies forming the highest beds of the Blue Anchor Formation, at Blue Anchor Point, Somerset. The tooth, now lost, may have been a broken haramiyid tooth (Owen 1871), but its relatively large size (c. 4·4 mm) suggested to Clemens *et al.* (1979) that it may have been a tritylodontid or similar therapsid.

16. TRACE FOSSILS

by ANDREW SWIFT and CHRISTOPHER J. DUFFIN

Trace fossils from the Penarth Group, despite their abundance, have been largely overlooked and have received very little taxonomic attention. Only forms from the Westbury Formation have received detailed study, yet biogenic structures also occur commonly in both members of the Lilstock Formation. At places the sediments of the Langport Member especially were heavily bioturbated, which may be related to opportunistic animals re-colonizing areas after storm generated or seismically triggered redeposition of sediments, evidence for which is abundant in the rocks of the Langport Member (Hallam 1960; Swift 1995a).

The trace fossils of the Penarth Group record a wide range of behavioural patterns made by aquatic animals and are especially varied in the harder, sandier beds in the Westbury Formation. From these traces it can be deduced that many organisms which have left little or no evidence in the form of body fossils were in fact important members of the Penarth Group biota. A good example is the wide variety of traces made by xiphosurid crustaceans (Ager and Edwards 1986; Wang 1993), which are unrecorded as body fossils. Also, many dwelling and/or feeding burrows are attributable to marine worms, which again have a very poor record as body fossils. One of few trace fossils which undoubtedly illustrates the behaviour of animals well-represented in the body fossil record is the bivalve resting trace known as *Lockeia* (Pl. 31, fig. 1).

A classification of trace fossils based on the behaviour which the trace is believed to record was devised by Seilacher (1953a, 1953b, 1964). The scheme was later refined by Bromley (1996). The classes are Cubichnia (resting traces), Repichnia (crawling traces), Pascichnia (grazing traces), Fodinichnia (feeding traces), Domichnia (dwelling traces), Agrichnia (gardening traces and traps), Praedichnia (predation traces), Equilibrichnia (equilibrium traces, i.e. animals adjusting to the level of the sediment water interface), Fugichnia (escape traces), Aedificichnia (edifices constructed above the substrate) and Calichnia (structures made for breeding purposes). Some indication of which behavioural class is appropriate for each Penarth Group trace fossil is given in the systematic remarks. Seilacher (e.g. 1964, 1967) also introduced the concept of sediment- and environment-related trace fossil associations and delineated a number of 'ichnofacies', characterized by certain forms which define a particular depth-zone or point on a shore to basin transect. All of the trace fossils identified from the Penarth Group are constituents of either the *Skolithos* or the *Cruziana* ichnofacies of Seilacher (1964, 1967). Both often indicate a nearshore to shelf environment; more specifically the *Skolithos* association is commonly developed in a moderate to high energy environment in a lower littoral to infralittoral setting where sediments are regularly disturbed and subject to sudden erosion or deposition. The animals which created the traces of the *Cruziana* ichnofacies most probably inhabited a more basinward environment, but still in shallow water, between daily wavebase and storm wavebase. Thus a shallow water origin, at least for those sediments of the Penarth Group which preserve trace fossils, is strongly suggested.

Only two studies have dealt with Penarth Group trace fossils in formally documented and illustrated case studies. Both focused on trace fossils from the Westbury Formation or equivalent strata. The first was by Wright and Benton (1987), who described traces from shallow-water shore-face sandstone facies coeval with the Westbury Formation in an Upper Triassic outlier in the Needwood Basin, Staffordshire. An extensive ichnofauna was reported, dominated by *Pelecypodichnus* (=*Lockeia*), but also including *Arenicolites*, *Kouphichnium*, *Palaeophycus*, *Planolites* and *Rusophycus*, plus other trails and burrows. This varied ichnofauna records the activities of a diverse suite of organisms, some of which moved over the sand leaving groove and ridge trails, i.e. *Kouphichnium*, whilst others lived within the sand leaving the trace *Arenicolites* and bivalve escape burrows, or lived temporarily at the lower surface along the sand-clay interface, i.e. *Palaeophycus*, *Planolites*, *Rusophycus* and small trails. Body fossils are represented by abundant moulds of the bivalve *Isocyprina* (*Eotrapezium*) sp., which was probably responsible for the majority of *Lockeia* traces.

The other substantial work on Penarth Group trace fossils was by Wang (1993), who discussed the provenance of an impressive trace fossil assemblage preserved in thick Westbury Formation sandstones at Garden Cliff, Westbury, with some reference to similar occurrences at Patchway, near Bristol. He concentrated on traces which could be referred to the activities of xiphosurid arthropods and discussed the affinities of two patterns of scratch marks, three types of lunate marks and a bilobate furrow. Wang (1993) noted that the traces bore most resemblance to the ichnogenera *Cruziana*, *Rusophycus*, *Selenichnites*, *Kouphichnium* and *Diplichnites*, but he made definite assignments only to *Cruziana perucca* Seilacher and *Selenichnites* isp. The traces were thought to have been made after storm events, when fresh oxygenated conditions permitted xiphosurid activity. Wang (1993) also recorded the sponge-like trace fossil *Spongeliomorpha*, which is thought to represent the dwelling burrow-system of an arthropod. Also identified were *Diplocraterion parallelum*, *Skolithos* isp. and *Lockeia* isp., which are usually considered to be of non-arthropod origin.

In a short abstract Magor (1978) reported trace fossils from the sandstones of Garden Cliff, suggesting that the traces recorded the foregathering of arthropods for the purpose of egg-laying and feeding.

Upper Triassic trace fossils from the Williton Member of the Blue Anchor Formation, which underlies the Penarth Group in Somerset, were reported by Mayall (1981). Recorded genera include *Diplocraterion*, *Planolites*, *Muensteria*, *Rhizocorallium*, *Siphonites* and *Arenicolites*, the last being the only trace fossil recorded from the lower Rydon Member of the Blue Anchor Formation. Although these trace fossils are not from the Penarth Group, they are associated with the onset of the transgression which led to the deposition of the Penarth Group, and as such are dealt with here.

An additional genus is recorded here for the first time. This is *Asteriacites*, a star-shaped trace attributable to a five armed echinoderm (Pl. 31, fig. 2). In this case the maker was an ophiuroid ('brittle-star'), rather than an asteroid ('star-fish'). Representatives of the former are fairly common as body fossils in the Westbury Formation, whilst the latter are so far unrecorded in the Penarth Group. In addition, the morphology of the *Asteriacites* specimens indicates long, thin arms originating quite discretely from a central disc, which is the characteristic body plan of ophiuroids. The two specimens which have been found came from the Westbury Formation at Aust Cliff.

No trace fossils made by terrestrial animals, in the form of locomotion marks such as footprints, have been recorded from the Penarth Group.

Text-figure 31 contains generalized sketches of typical trace fossils recorded from the Penarth Group and some representative 'taxa' are figured on Plate 31. Specimens illustrated in Plate 31, figures 1, 7–10 are held in the A. Swift collection in the Geology Department, University of Leicester (prefix BC); those in Plate 31, figures 5–6 are held in The New Walk Museum, Leicester (prefix NW); that in Plate 31, figure 4 is held in the Natural History Museum, London (prefix V) and that in Plate 31, figure 2 is held in the Bristol City Museum and Art Gallery (prefix Cc). Plate 31, figure 3 was photographed in the field at Bantycock Quarry, Newark, as it was too friable for removal. Specimens shown on Text-figure 32A–B are held in the Bristol City Museum and Art Gallery (prefix Cb) and C–D are held in The Natural History Museum, London (prefix P).

SYSTEMATIC REMARKS

Only brief systematic notes are given for each of the 'taxa' identified from the Penarth Group because of the many unresolved problems attending the taxonomic treatment of trace fossils, i.e. should they be classified according to form taxonomy or should the various traces made by one identified organism be linked in a truly biological classification? Even without taxonomic problems, the identification of trace fossils is fraught with difficulty, since many are 'overwritten' by subsequent activity and mimicked by the quite separate activities of unrelated taxa. For these reasons, there is no established hierarchy and the classification of trace fossils usually begins and ends at generic level.

Basic information for this section has been drawn from Häntzschel (1975) and Bromley (1996). All but one of the genera discussed have previously been recorded only from the higher beds of the Blue Anchor Formation and Westbury Formation, but many are also present in the higher beds of the Penarth Group. The non-marine beds of the Cotham Member contain few unequivocal trace fossils, but the soles of ripples in the unit which were formed in an estuarine environment contain traces which are similar to some of those below (Pl. 31, figs 8–10).

Genus SKOLITHOS Haldeman, 1840
Text-figure 31A

Remarks. Simple unbranched perpendicular tube. Usually found *en masse*. Identified as the dwelling trace of suspension feeders such as worms or phoronids.

Genus ARENICOLITES Salter, 1857
Plate 31, figure 3; Text-figure 31B

Remarks. Simple U-tube without spreiten. Represents the burrow of suspension feeding worms. Classed as a dwelling trace.

Genus DIPLOCRATERION Torell, 1870
Text-figure 31C

Remarks. Simple U-tube with spreiten situated vertical to the bedding plane with two parallel limbs. It is an example of an equilibrium trace i.e. an animal adjusting for being covered or uncovered by sediment (hence the development of spreiten).

Genus RHIZOCORALLIUM Zenker, 1836
Text-figure 31D

Remarks. Similar to *Diplocraterion* but the limbs of the burrow curve in the top part and are somewhat oblique to the bedding plane. This is also a burrow of a deposit feeding animal. As well as records from the Westbury Formation, this ichnogenus has also been recorded from the Langport Member of the Devon coast by Hallam (1960).

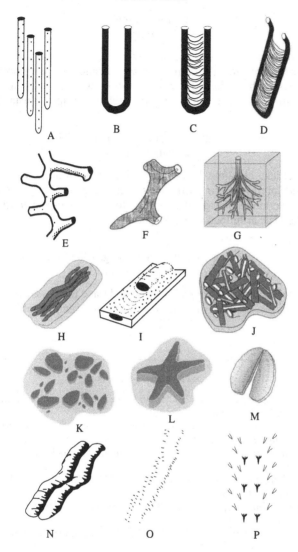

TEXT-FIG. 31. Frequently encountered trace fossils of the Penarth Group. A, *Skolithos* Haldeman, 1840. B, *Arenicolites* Salter, 1857. C, *Diplocraterion* Torell, 1870. D, *Rhizocorallium* Zenker, 1836. E, *Thalassinoides* Ehrenberg, 1944. F, *Spongeliomorpha* De Saporta, 1887. G, *Chondrites* Sternberg, 1833. H, *Planolites* Nicholson, 1873. I, *Muensteria* Sternberg, 1833. J, *Siphonites* De Saporta, 1872. K, *Lockeia* James, 1879. L, *Asteriacites* von Schlotheim, 1820. M, *Rusophycus* Hall, 1852. N, *Cruziana* d'Orbigny, 1842. O, *Diplichnites* Dawson, 1873. P, *Kouphichnium* Nopcsa, 1923. Not to scale.

Genus THALASSINOIDES Ehrenberg, 1944
Text-figure 31E

Remarks. Many of the branching burrow systems found in the Penarth Group are attributed to this familiar genus. Branches are smooth and unornamented with Y-shaped forkings, often with inflated areas at intervals within the system. Most are attributable to decapod crustaceans.

Genus SPONGELIOMORPHA De Saporta, 1887
Text-figure 31F

Remarks. Thick, elongate bodies reminiscent of antlers, with ramifications and lateral tapering offshoots. Surface bears scratches intersecting at acute angles. Originally thought to be sponges but most often now considered as arthropod dwelling burrows.

Genus CHONDRITES Sternberg, 1833
Plate 31, figure 4; Text-figure 31G

Remarks. One of the most frequently recorded trace fossils, *Chondrites* is a complex burrow system of ramifying branches running off a single

---

EXPLANATION OF PLATE 31

Fig. 1. *Lockeia* James, 1879; AS BC 97/W/3; Westbury Formation; Bantycock Quarry, Newark, Nottinghamshire; ×1·5.

Fig. 2. *Asteriacites* von Schlotheim, 1820; BCMAG Cc 476; Westbury Formation; Aust Cliff, Gloucestershire; ×1.

Fig. 3. *Arenicolites* Salter, 1857; Westbury Formation; Bantycock Quarry, Newark, Nottinghamshire; ×0·25.

Fig. 4. *Chondrites* Sternberg, 1833; BMNH V 13168; 'White Lias' (Langport Member); Lavernock, South Glamorgan; ×1.

Fig. 5. Type 3 coprolite; NW 326'1889; Westbury Formation; Aust Cliff, Gloucestershire; ×1·5.

Fig. 6. *Planolites* Nicholson, 1873; NW 33'1921/7; Westbury Formation; Glen Parva, Leicester; ×2.

Fig. 7. cf. *Palaeophycus* Hall, 1847; AS BC 97/W/6; Westbury Formation; Bantycock Quarry, Newark, Nottinghamshire; ×0·5.

Fig. 8. Meandering and crossing grazing trails; AS BC 97/C/2; Cotham Member; Bantycock Quarry, Newark, Nottinghamshire; ×1·5.

Fig. 9. Indeterminate resting/grazing traces; AS BC 97/C/1; Cotham Member; Bantycock Quarry, Newark, Nottinghamshire; ×1·5.

Fig. 10. Indeterminate grazing/resting traces; AS BC 97/C/5; Cotham Member; Bantycock Quarry, Newark, Nottinghamshire; ×1·5.

PLATE 31

'entry hole' at the sediment/water interface. In three dimensions the shape usually resembles a many-branched plant or tree with branches of equal width throughout which do not cross each other. Sometimes faecal pellets are present within the system. The animal which produced *Chondrites* is thought to be a chemosymbiont i.e. one which employs bacteria as co-operative organisms to ameliorate poisonous conditions. As such it may be that the trace could in part be thought of as a gardening trace (agrichnia), but elements of feeding and dwelling also play a part.

Genus PLANOLITES Nicholson, 1873
Plate 31, figure 6; Text-figure 31H

Remarks. A straight to gently curved, non-branching, cylindrical grazing trail which proceeds generally horizontal to bedding, made by a deposit feeder. The fill is structureless but usually differs in colour from the surrounding sediment.

Genus PALAEOPHYCUS Hall, 1847
cf. Plate 31, figure 7

Remarks. Similar to, but differs from *Planolites* in that it tends to branch, has a similar infill to the surrounding sediment and proceeds obliquely to the bedding plane. Also more sinuous. Details of the wall material indicate that *Palaeophycus* might represent a permanent burrow inhabited by a suspension feeder, carnivore or scavenger. As such it would be classed as a dwelling trace.

Genus MUENSTERIA Sternberg, 1833
Text-figure 31I

Remarks. Somewhat similar to *Planolites* and also made by deposit feeders, but the infill has transverse annulations, and the trail tends to be branched. It is a so-called 'stuffed-burrow' where the laminated structure is caused by back-filling. A closely-related genus, *Taenidium* Heer, tends to be smaller.

Genus SIPHONITES De Saporta, 1872
Text-figure 31J

Remarks. Simple tubes usually about 10 mm in diameter, several tens of millimetres long, with a sandy lining, mostly washed out and collapsed horizontally and randomly onto a bedding plane. Rather an insecure genus, which may simply represent the post-disturbance remnants of broken-up cylindrical burrows such as *Skolithos*, *Arenicolites*, etc.

Genus LOCKEIA James, 1879
Plate 31, figure 1; Text-figure 31κ

Remarks. This is the compression resting trace of a small, burrowing bivalve and takes the form of a small-medium, variable, usually symmetrical, broadly almond-shaped mass which draws to a point at each end. Usually occurs in numbers on bedding planes in the Westbury Formation, indicating communal activity. *Pelecypodichnus* Seilacher is a junior synonym.

Genus ASTERIACITES von Schlotheim, 1820
Plate 31, figure 2; Text-figure 31ʟ

Remarks. One of the few trace fossils where the trace-maker is in no doubt. The outline mimics well the five-fold symmetry of asteroid or ophiuroid echinoderms, and represents the trace of the resting or predation behaviour (certain echinoderms are known to 'burrow' into sediment for buried prey) of these animals. The figured examples from the Penarth Group were made by an ophiuroid.

Genus RUSOPHYCUS Hall, 1852
Text-figure 31м

Remarks. This is a resting trace, again thought to be that of an arthropod. The trace is paired and takes the form of broad, short furrows like coffee beans, ornamented with transverse 'wrinkles' or ridges and with a groove between.

Genus CRUZIANA d'Orbigny, 1842
Text-figure 31ɴ

Remarks. This is a crawling trace comprising roughly parallel, double burrow-like trackways, ornamented with fine oblique cross ribs. Most often associated with trilobite activity, but also applied to traces made by other arthropods.

Genus DIPLICHNITES Dawson, 1873
Text-figure 31o

Remarks. An obvious crawling trace, recording the light impressions of claws or appendages, probably of an arthropod, in two parallel series of fine ridges, offset obliquely to the track axis.

Genus KOUPHICHNIUM Nopcsa, 1923
Text-figure 31P

Remarks. Another track-way recording the progress of a crawling arthropod, but in this case the marks are more complex and variable and were produced by a combination of legs, appendages and claws, reminiscent of the impressions of bird feet.

Genus SELENICHNITES Romano and Whyte, 1990

Remarks. Rather poorly defined trilobate trace which preserves the crude outline of an arthropod partially buried in the sediment. As such it would be classed as a resting or possibly feeding trace.

COPROLITES

'Coprolite' is a term introduced by Buckland (1829) for fossilized faeces. He first recognized coprolites from the Westbury Formation bone bed at Aust Cliff when he wrote, 'the small black bodies of various shapes, and having a polished surface, which occur mixt [sic] with bones in the lowest strata of the lias on the banks of the Severn, near Bristol, are also of faecal origin: – they appear to be co-extensive with this bone bed, and occur at many and distant localities' (Buckland delivered this paper on May 1st, 1829, but it was not published until 1834, see p. 142). He originally proposed the name *Nigrum graecum* for these fossil faeces, surmising that they may have been produced by reptiles or fishes. *Nigrum graecum* is not a valid ichno-taxon, as it lacks illustration and description.

Coprolites with an internal spiral structure have attracted much attention and have been variously interpreted as faecal in origin (McAllister 1985), enterospirae (fossilized intestinal spiral valves: Neumayer 1904; Fritsch 1907; Williams 1972; Duffin 1979) or cololites (colon contents *sensu* Storrs 1994). The term 'coprolite' is used informally here to encompass both intestinal tract contents and expelled material.

The coprolite assemblage of the Westbury Formation bone beds was described by Duffin (1979), who ordered the structures into four broad morphological types and discussed the possible origins of each group.

Type 1 coprolites (Text-fig. 32A)
These are large (up to 80 mm long), usually brown, often tapered coprolites with a well-defined spiral structure. The spiral folds are clearly visible on the external surface and reflect an internal arrangement of overlapping folds. The spiral folds tend to be evenly distributed along the length of the specimen (the so-called 'amphipolar arrangement' of Neumayer 1904). Undigested vertebrate fragments contained within the ground mass may

TEXT-FIG. 32. Coprolite morphologies from the basal bone bed of the Westbury Formation at Aust Cliff, Gloucestershire. A, type 1 coprolite; BCMAG 4891; ×1. B, type 2 coprolite; BCMAG Cb 4898; ×2. C, type 3 coprolite; BMNH P 23353; ×2. D, type 4 coprolite; BMNH P 23353; ×2 (after Duffin 1979, pl. 21, figs 1, 8, 12; pl. 22, fig. 2 respectively).

include a variety of fish scales which are often arranged tangential or normal to the spiral folds. Discrete food boli may be apparent in thin section, accentuated by changes in colour and density of the groundmass. Coprolites of this type may also contain crustacean remains (e.g. *Tropifer laevis* and possible isopods, see Chapter 9).

The spiral structure of these coprolites indicates that the producer must have possessed a spiral valve in the intestinal tract. Penarth Group vertebrates which contain such a structure for increasing the surface area for absorption of the end-products of digestion include the chondrichthyans, palaeoniscid chondrosteans and dipnoans. The presence of undigested vertebrate and rarer invertebrate food material in the coprolites suggests a predatory or scavenging mode of life for the coprolite-producer. Thus, the most likely candidates for the production of Type 1 coprolites are the sharks, possibly the myriacanthid holocephalans, and the palaeoniscid chondrosteans. Dipnoans may have produced those coprolites which contain arthropods.

Type 2 coprolites (Text-fig. 32B)
These light brown to black, elongate (*c*. 30 mm long) coprolites also have a well-defined spiral structure. They may be amphipolar, or in some cases have the spiral coils concentrated towards one end (the blunt end) of the specimen ('heteropolar type' of Neumayer 1904). Visible undigested vertebrate and invertebrate inclusions are lacking. The combination of spiral structure and lack of inclusions suggests that these coprolites may have been produced by either the dipnoan *Ceratodus* or possibly even the myriacanthid holocephalans, which are believed to have been opportunistic feeders.

Type 3 coprolites (Pl. 31, fig. 5; Text-fig. 32C)
Coprolites of this type are usually small (maximum 30 mm long), capsule-shaped bodies lacking both spiral structure and vertebrate and invertebrate inclusions. They are often black and shiny due to polishing and abrasion during post-fossilization transport. They are usually homogenous in thin section and often contain disseminated pyrite. The producer this type of coprolite has not been identified.

Type 4 coprolites (Text-fig. 32D)
These are flattened, shiny black forms up to 30 mm long. Included undigested scales and teeth may occasionally break the surface of the specimen, but there is no internal spiral structure. It is unclear which animal was responsible for this coprolite type, although small reptiles are one possibility.

17. PLANTS

by ANDREW SWIFT

Macerated and indeterminate plant debris and woody matter is distributed widely through the sediments of the Penarth Group and can be found in all three units, but determinable plant fossils are largely confined to the Cotham Member, and to particular horizons within that unit. The source of almost all this plant material is the aquatic liverwort, *Naiadita lanceolata*, whose remains occur in dense tangled mats on certain bedding planes (Pl. 32, figs 1–4). Preservation can be excellent and fine details discernible, which allowed Harris (1931, 1937, 1938, 1939) to produce meticulous and thorough studies to which little can be added. The reader is referred to these texts for a full description of this unusual plant, which shows an intriguing mixture of characters. Sometimes associated with *Naiadita* in its occurrences in and around Bristol is the much rarer liverwort, *Hepaticites solenotus* Harris, which appears to have been recorded only by its author (see Harris 1938). Preservation of *H. solenotus* is generally poor and its affinities uncertain. Plate 32, figures 3–4 illustrate the surface of a block which apparently contains remains of *H. solenotus* (Harris 1938, p. 64), but little can be accurately discerned.

Since the time of Harris's work with *Naiadita* very little new data has come to light regarding plants from the Penarth Group. However, shortly before his death, Harris wrote some notes on a plant from sandstones probably referable to the Quarella Sandstone, a littoral Rhaetian facies in South Wales. This local development is probably equivalent to some portion of the Westbury Formation. He established that the plant was a fern and identified it as *Clathropteris meniscoides* Brongniart. His notes and two illustrations were published by Ager and Edwards (1986), but as far as is known the plant has not been formally described.

Other recorded plant fossils are few, but leaves of the bennettitalian *Otozamites bechei* Brongniart have been recorded from Waterloo Glen near Larne, Northern Ireland and from near Cowbridge, South Wales (Pl. 32, fig. 6), in deposits which are equivalent to some part of the Penarth Group (Harris and Jackson 1961). In the collections of The Natural History Museum in London is a specimen of *Otozamites obtusus* (Lindley and Hutton) from the 'sun bed' at the top of the Langport Member at a quarry near Radstock, Somerset (Pl. 32, fig. 5). A cast of a stem apparently related to plants such as *Neocalamites* or *Equisetites* from the 'Rhaetic' at Feckenham, Worcestershire is also held in The Natural History Museum collections and is illustrated on Plate 32, figure 8. Both of these latter two specimens were mentioned by Harris (1938, p. 72), who also recorded '?*Otozamites* sp.' and 'seeds' from marginal sandstones at St Hilary, Glamorgan and lignite from a Westbury Formation bone bed at Charlton Mackrell, Somerset. 'Branched filaments resembling moss rhizoids' were also noted by Harris (1938, p. 71), but these fragments cannot be further elucidated.

Figured specimens are held in The Natural History Museum, London (Pl. 32, figs 1–8) and the A. Swift collection, Geology Department, University of Leicester (Pl. 32, fig. 9).

SPECIES DESCRIPTIONS

Division BRYOPHYTA Schimper, 1879
Class HEPATICOPSIDA Rothmaler, 1951
Order SPHAEROCARPALES Cavers, 1910?
Genus NAIADITA Buckman, 1850
Naiadita lanceolata Buckman, 1850 emend. Harris, 1938
Plate 32, figures 1–4

Description. Stem slender. Rhizoids are trains of cells one cell thick with evenly thickened walls. Leaves unicellular, arranged spirally, attached transversely by broad base, lacking midrib. Archegonia lateral on short stalks which elongate to become fruit pedicels; archegonia originally naked, later enclosed in a perianth resembling foliage leaves. Sporophyte consists of spiral capsule which develops in archegonium venter, becoming partly embedded in tissue of gametophyte pedicel. Capsule wall unicellular, ripe capsule full of spore-tetrads. (Adapted from the diagnosis of Harris 1938).

Remarks. The whole plant has a weak and spindly appearance, which supports the conjecture that *Naiadita lanceolata* was an aquatic plant, needing the support of water to remain upright. It is commonly associated with freshwater ostracods, occasional insect remains and rare *Euestheria*, but paradoxically sometimes also with marine fish and molluscs. Such associations suggest a complex environment of deposition; a marginal freshwater/brackish water system periodically invaded by high tides or storm surges bringing in marine elements might be envisaged.

Conditions favouring the growth and preservation of *N. lanceolata* occurred rarely during deposition of the Cotham Member, and not at all during the deposition of the remainder of the Penarth Group. In addition, there is a relatively small area of distribution. The most prolific and best preserved material occurs in exposures in and around Bristol (e.g. Short 1904). *N. lanceolata* is rare in the more marine influenced Cotham Member deposits of Dorset, west Somerset and South Wales, and also is rarely recorded from the Cotham Member outcrop north of Warwickshire, despite the greater thickness and non-marine nature of the Cotham Member in its more northerly developments.

For many years the peculiar mixture of characters led to confusion and debate over the affinities of *N. lanceolata* and a relationship to the mosses has been proposed from time to time (Gardner 1886; Seward 1904). Sollas (1901) believed it to be a lycopod and assigned it to the genus *Lycopodites*. It wasn't until Harris's work (1938, 1939) that the questions were satisfactorily resolved. Hemsley (1989) studied the ultrastructure of the spore wall of *N. lanceolata* and was able to confirm the mix of characters which make the plant something of a chimaera. He concluded that assignment to any particular group of liverworts was not possible, but that a close relationship was demonstrated with the sphaerocarpales and marchantiales.

Distribution. Cotham Member at many localities in east Somerset and Gloucestershire. Rare farther north.

Order *Incertae sedis*
Genus HEPATICITES Walton, 1925
Hepaticites solenotus Harris, 1938
Plate 32, figures 3–4

Remarks. The material of *H. solenotus* is fragmentary and poorly preserved. Harris (1938) managed to discriminate most parts of the plant, collate a diagnosis and give drawings of examples of the thallus and gemmae, but detail is obscure. The plant is associated in its few occurrences with *Naiaidita*, and seems to be mainly distinguishable by its colour, appearing light orange-brown in oil.

Division GYMNOSPERMOPHYTA
Class CYCADOPSIDA Barnard and Long, 1975
Order CYCADALES Engler, 1892
Family CYCADACEAE Persoon, 1807
Genus OTOZAMITES Braun, 1843
Otozamites bechei Brongniart, 1849
Plate 32, figure 6

Description. Pinnae entire, elongate, rounded at termination, length to width ratio 5:1. Margins of pinnae flat, not thickened, surfaces smooth. Apices of pinnae obtuse, overlapping. Veins slender, numerous, traversing laminae at concentration of 50–60 per 10 mm, diverging from pinna base, forking freely. Epidermal cells produce fine markings on lower pinnae, causing faint protruberances on upper surface but more distinctive bulging on lower surface. Stomata consist of two rows of faint pits between pairs of veins. (Adapted from the diagnosis of Harris and Jackson 1961).

PLATE 32

Remarks. *Otozamites bechei* is broadly similar in morphology to certain species of the familiar Upper Carboniferous genus *Alethopteris*. Its occurrences appear to be limited to marine facies deposited close to shorelines, as in the marginal deposits around Cowbridge in South Wales, where some leaves evidently survived the short distance from their terrestrial habitat to an eventual repository in marine sediment.

A specimen of a closely similar species, *O. obtusus* (Lindley and Hutton), was recovered the Langport Member near Radstock (Pl. 32, fig. 5). This occurrence has not been previously detailed or described, apart from a brief mention in Harris (1938). It may be that the record of *O. obtusus* in Kellaway and Welch (1993, p. 142) refers to the same occurrence.

Distribution. Known from deposits coeval with the Penarth Group in South Wales (Cowbridge) and Northern Ireland (Waterloo Bay, Larne). Also a single record of *Otozamites obtusus* from the top of the Langport Member near Radstock, Somerset.

'ALGAE' (PLANT PROTISTS)

Fossils attributed to certain types of algae have been reported sporadically from the Penarth Group and Harris (1938) gave detailed descriptions of two species from the Cotham Member, associated with *Naiadita*. His specimens have been re-examined and proved to be extremely small and indistinct. Harris (1938, p. 11) stated that sometimes these algae are

---

EXPLANATION OF PLATE 32

Figs 1–2. *Naiadita lanceolata* (Buckman, 1850); BMNH V 4015c; Cotham Member; Redlands, Bristol. 1, shoots, mostly complete; 2, two individual shoots. Both ×2.

Figs 3–4. *Naiadita lanceolata* (Buckman, 1850) and *Hepaticites solenotus* Harris, 1938; BMNH In 27082; Cotham Member; Coldharbour Farm, Redlands, Bristol. Both ×4.

Fig. 5. *Otozamites obtusus* (Lindley and Hutton); BMNH V 10316; 'Sun Bed' (top of Langport Member); quarry above Ludlow's Quarry, Radstock, Somerset; ×1.

Fig. 6. *Otozamites bechei* Brongniart, 1849; BMNH V 42427; 'Rhaetic horizon 3 of Strahan' (see Strahan and Cantrill 1904); Cowbridge railway cutting, Glamorgan; ×1.

Fig. 7. Indeterminate fragment of wood; BMNH V 10388; 'Rhaetic'?; Wilsbridge, near Keynsham, Bristol; ×1.

Fig. 8. '*Neocalamites* or *Equisetites*'; BMNH V 20587; 'Rhaetic'; Lane House, Feckenham, Worcestershire; ×1.

Fig. 9. Cotham or Landscape Marble; AS 11/24A; Cotham Member; Cotham, Bristol; ×0·67.

revealed only by chemical treatment. In view of this and the fact that they are undetectable in hand specimen it is felt inappropriate to do more than record them in this field guide. The two described species are *Botryococcus braunii* Kützing, and *Stenixys cosmaroides* Harris (Harris 1938). Filamentous and other algal remains are fairly common in palynological preparations (e.g. Warrington 1982) and two species of coccolith, the microscopic plates built by certain types of algae, also occur in very small numbers through the Penarth Group around Watchet (Hamilton 1982).

No other species of algae have been reported, but peculiar structures built by unspecified algae are preserved in the distinctive Cotham or Landscape Marble, a stromatolitic flinty limestone which occurs around the top of the Cotham Member (Pl. 32, fig. 9, see also Frontispiece and Text-fig. 2). The aborescent patterns capped by 'atmospheres' which mimic treelines and skyscapes are very appealing and have led to the stone being used ornamentally, especially around the turn of the nineteenth century. The structures were identified as algal tufts by Mayall and Wright (1981). The Cotham Marble has a patchy outcrop and like *Naiadita*, tends to be best developed in east Somerset, the Bristol area and Gloucestershire.

REFERENCES

ADAMS, H. and ADAMS, A. 1853–58. *The genera of Recent Mollusca arranged according to their organisation*. Vol. 1 (1853–54), 484 pp.; Vol. 2 (1854–58), 661 pp.; Vol. 3 (1858), plates (136). J. van Voorst, London.

AGASSIZ, L. J. R. 1832. Untersuchungen über die fossilien Fische der Lias-Formation. *Neues Jahrbuch für Mineralogie, Geologie und Palaeontologie*, 139–149.

—— 1833–44. *Recherches sur les poissons fossiles*. 5 vols. Neuchatel, 1420 pp.+supplement.

—— 1840–45. *Études critiques sur les Mollusques fossiles*; 1st part 1840; 2nd and 3rd parts 1842 as '*Monographie des Myes du Jura et de la Craie Suisse*'. Neuchâtel, xxii+287 pp. Appendices, 1845.

—— 1844. Synoptic table of British fossil fishes, arranged in the order of the geological formations. *Report of the British Association for the Advancement of Science*, **1843**, 194–207.

—— and DESOR, P. J. E. 1847. *Catalogue raisonné des familles, des genres et des espèces de la classe des Echinodermes*. Paris, 167 pp.

AGER, D. V. and EDWARDS, D. 1986. The fauna and flora of the Rhaetian of South Wales and adjacent areas. *Nature in Wales*, **4**, 71–79.

AIGNER, T. 1982. Calcareous tempestites: storm dominated stratification in Upper Muschelkalk limestones (Middle Trias, S.W.-Germany). 180–198. *In* EINSELE, G. and SEILACHER, A. (eds). *Cyclic and event stratification*. Springer-Verlag, Berlin, xiv+536 pp.

AINSWORTH, N. R. 1989a. Rhaetian, Hettangian and Sinemurian Ostracoda from the Fastnet Basin, offshore southwest Ireland. *Bulletin of the Geological Survey of Ireland*, **1989/4**, 107–150.

—— 1989b. Systematic description of new species. *In* RUTHERFORD, M. M. and AINSWORTH, N. R. Micropalaeontological and stratigraphical recognition of the Triassic/Jurassic boundary in the North Celtic Sea and Fastnet basins. 56–65. *In* BATTEN, D. J. and KEEN, M. C. (eds). *Northwest European micropalaeontology and palynology*. Ellis Horwood, Chichester, x+298 pp.

—— 1990. Uppermost Rhaetian to Lower Bajocian Ostracoda from the Porcupine, Slyne, Erris and Donegal basins, offshore west Ireland. *Bulletin of the Geological Survey of Ireland*, **1990/4**, 169–200.

—— and BOOMER, I. in press. The Triassic. *In* ATHERSUCH, J., KEEN, M. and WILKINSON, I. (eds). *A stratigraphical index of British Ostracoda*. 2nd edition. Chapman and Hall, London.

—— and HORTON, N. F. 1987. Mesozoic micropalaeontology of exploration well Elf 55/30–1 from the Fastnet Basin, offshore southwest Ireland. *Journal of Micropalaeontology*, **5**, 19–29.

—— O'NEILL, M. and RUTHERFORD, M. M. 1989. Jurassic and Upper Triassic biostratigraphy of the North Celtic Sea and Fastnet basins. 1–44. *In* BATTEN, D. J. and KEEN, M. C. (eds). *Northwest European micropalaeontology and palynology*. Ellis Horwood Ltd., Chichester, x+298 pp.

258   *Fossils of the Rhaetian Penarth Group*

—— —— —— CLAYTON, G., HORTON, N. F. and PENNEY, R. A. 1987. Biostratigraphy of the Lower Cretaceous, Jurassic and uppermost Triassic of the North Celtic Sea and Fastnet basins. 611–622. *In* BROOKS, J. and GLENNIE, K. (eds). *Petroleum geology of north west Europe*. 2 vols. Graham and Trotman, London.

ALBERTI, F. A. von 1832. Über bunten Sandstein, Muschelkalk und Keuper. *Jahrbuch für Mineralogie, Geognosie, Geologie und Petrefaktenkunde*, Heidelberg, 226–227.

—— 1834. *Beitrag zu einer Monographie des bunten Sandsteins, Muschelkalks und Keupers, und die Verbindung dieser Gebilde zu einer Formation.* Stuttgart and Tübingen, xx+366 pp.

ALDRIDGE, R. J. and PURNELL, M. A. 1996. The conodont controversies. *Trends in Ecology and Evolution*, 11, 463–468.

ALEXANDER, C. I. 1933. Shell structure of the ostracode genus *Cytheropteron*, and fossil species from the Cretaceous of Texas. *Journal of Paleontology*, 7, 181–214.

ALLEN, H. A. 1899. Catalogue of types and figured specimens of British Gastropoda and Scaphopoda from the Rhaetic Beds, Lias and Inferior Oolite, preserved in the Museum of Practical Geology, London. *Summary of Progress of the Geological Survey of Great Britain for* 1898, 217–228.

—— 1904. Catalogue of types and figured specimens of British lamellibranchiata from the Rhaetic Beds and Lias, preserved in the Museum of Practical Geology, London. *Summary of Progress of the Geological Survey of Great Britain for* 1903, 172–177.

AMMON, L. von 1878. Die Gastropoden des Hauptdolomits und Plattenkalkes de Alpen. *Abhandlungen des zoologisch-mineralogischen Vereins zu Regensburg*, 11, 1–72.

ANDERSON, F. W. 1964. Rhaetic Ostracoda. *Bulletin of the Geological Survey of Great Britain*, 21, 133–174.

ANDREAE, A. 1887. Die Glossophoren des Terrain à chailles der Pfirt. *Abhandlungen zur geologischen Specialkarte von Elsass-Lothringen*, Band 4, Heft 3, 1–45.

ANDREWS, C. W. 1922. Description of a new plesiosaur from the Weald Clay of Berwick (Sussex). *Quarterly Journal of the Geological Society, London*, 78, 285–295.

ANTIA, D. D. J. 1979. Bone-beds: a review of their classification, occurrence, genesis, diagenesis, geochemistry, palaeoecology, weathering and microbiotas. *Mercian Geologist*, 7, 93–174.

—— and SYKES, J. H. 1979. The surface texture of quartz grains from a Rhaetian bone bed, Blue Anchor Bay, Somerset. *Mercian Geologist*, 7, 205–210.

APOSTOLESCU, V. 1959. Ostracodes du Lias du Bassin de Paris. *Révue de l'Institut Français du Petrole et Annales des Combustibles Liquides*, 14, 795–826.

ARKELL, W. J. 1930. The generic position and phylogeny of some Jurassic Arcidae. *Geological Magazine*, 67, 297–310, 337–352.

—— 1933. *The Jurassic System in Great Britain.* Clarendon Press, Oxford, xii +681 pp.

—— 1934. A monograph of British Corallian Lamellibranchia. *Monograph of the Palaeontographical Society*, 86 (389), 229–276, pls 29–36; 87 (393), 277–324, pls 37–44.

AUDOUIN, J. V. and EDWARDS, H. M. 1832–34. *Recherches pour servir à l'histoire naturelle du littoral de la France.* 2 vols. Paris.

AVNIMELECH, M. 1952. Revision of the tubular Monothalamia. *Contributions to the Cushman Foundation for Foraminiferal Research*, **3**, 60–68.

BAILEY, E. B. and ANDERSON, E. M. 1925. The geology of Staffa, Iona, and western Mull. *Memoirs of the Geological Survey of Great Britain*, 1–107.

BAIRD, W. 1849. Monograph of the family Limnadiadae, a family of entomostracous Crustacea. *Proceedings of the Zoological Society, London*, **17**, 84–90.

—— 1850. Description of several new species of Entomostraca. *Proceedings of the Zoological Society, London*, **18**, 254–257.

BANNER, F. T., BROOKS, M. and WILLIAMS, E. 1971. The geology of the approaches to Barry, Glamorgan. *Proceedings of the Geologists' Association*, **82**, 231–247.

BARNARD, P. D. W. and LONG, A. G. 1975. *Triradioxylon*—a new genus of Lower Carboniferous petrified stems and petioles together with a review of the classification of early Pterophytina. *Transactions of the Royal Society of Edinburgh*, **69**, 231–250.

BARNARD, T. 1950. Foraminifera from the Lower Lias of the Dorset coast. *Quarterly Journal of the Geological Society, London*, **105**, 347–391.

—— 1956. Some Lingulinae from the Lias of England. *Micropaleontology*, **2**, 271–282.

BARTENSTEIN, H. 1962. Neue Foraminiferen aus Unterkreide und Oberkeuper NW Deutschlands und der Schweiz. *Senckenbergiana Lethaea*, **43**, 135–149.

BARTRAM, A. W. H. 1977. The Macrosemiidae, a Mesozoic family of holostean fishes. *Bulletin of the British Museum (Natural History), Geology Series*, **29**, 137–234.

BATE, C. S. 1888. Report on the Crustacea Macrura collected by H.M.S. Challenger during the years 1873–1876. *Report of the Scientific results of the voyage of HMS Challenger, Zoology*, **24**, 1–942.

BATE, R. H. 1963. Middle Jurassic Ostracoda from north Lincolnshire. *Bulletin of the British Museum (Natural History), Geology Series*, **8**, 173–219.

—— 1978a. The Trias. 175–188. *In* BATE, R. H. and ROBINSON, E. (eds). *A stratigraphical index of British Ostracoda.* Seel House Press, Liverpool, 538 pp.

—— 1978b. *In* POOLE, E. G. Stratigraphy of the Withycombe Farm Borehole near Banbury, Oxfordshire. *Bulletin of the Geological Survey of Great Britain*, **68**, 1–63.

BATESON, W. 1886. The ancestry of the Chordata. *Quarterly Journal of Microscopical Sciences*, **26**, 535–571.

BAYLE, É. 1878. Fossiles principaux des terrains. *Explication de la carte géologique de la France*, **4**, Atlas (1) + explanation.

BENTON, M. J. 1984. Small companions for early dinosaurs. *Nature*, **307**, 111–112.

—— 1986. The Late Triassic reptile *Teratosaurus*—a rauisuchian, not a dinosaur. *Palaeontology*, **29**, 293–301.

—— 1990. Mass extinction: events; end-Triassic. 194–198. *In* BRIGGS, D. E. G. and CROWTHER, P. R. (eds). *Palaeobiology: a synthesis.* Blackwell Scientific Publications on behalf of the Palaeontological Association, 583 pp.

—— and SPENCER, P. S. 1995. *Fossil reptiles of Great Britain.* Geological Conservation Review Series, Vol. 10. Chapman and Hall, London, 386 pp.

—— JUUL, L., STORRS, G. W. and GALTON, P. M. 1999. Anatomy and systematics of the prosauropod dinosaur *Thecodontosaurus antiquus* from the Upper Triassic of southwest England. *Journal of Vertebrate Paleontology*, **19**.

BERG, L. S. 1937. A classification of the fish-like vertebrates. *Bulletin of the Academy of Leningrad Biology*, **1937**, 1277–1280.

—— 1958. *System der rezenten und fossilen Fischartigen und Fische.* Translated from the 2nd Russian edition. Verhandlungen des Deutschen Wissenschaftlichen, xii+310 pp.

BERNER, R. A. 1984. Sedimentary pyrite formation: an update. *Geochimica et Cosmochimica Acta*, **48**, 605–617.

BERTHELIN, G. 1879. Foraminifères du Lias moyen de la Vendée. *Revue de la Magasin de Zoologie, Paris, Série 3*, **7**, 24–41.

BEURLEN, K. 1944. Beiträge zur Stammesgeschichte der Muscheln. *Sitzungsberichte der Bayerische Akademie der Wissenschaften, Mathematisch-physikalische Klasse*, **1–2**, 133–145.

BEYRICH, H. E. von 1845. *Protocardia*, eine neue Gattung fossilier Muscheln. *Zeitschrift für Malakozoologie*, **2**, 17.

BLAINVILLE, H. M. D. de 1825–1827. *Manuel de malacologie et de conchyliologie.* F. G. Levrault, Paris, viii+664 pp.

—— 1830. *Cours de physiologie générale et comparée, profesé à la faculté des sciences des Paris.* Vol. 1. Rouen and Paris.

—— 1834. *Manuel d'actinologie ou de zoophytologie.* F. G. Levrault, Paris, 694 pp.

—— 1835. Description de quelques espèces de reptiles de la Californie, précédée de l'analyse d'un système general d'Erpetologie et d'Amphibiologie. *Nouvelles annales du Muséum (national) d'histoire naturelle, Paris*, **4**, 233–296.

BLAKE, J. F. 1872. On the Infralias in Yorkshire. *Quarterly Journal of the Geological Society, London*, **28**, 132–146.

—— 1876. Crustacea, subclass Entomostraca. 429–435. *In* TATE, R. and BLAKE, J. F. *The Yorkshire Lias.* John Van Voorst, London, viii+475 pp.

BØGGILD, O. B. 1930. The shell structure of the mollusks. *Det Kongelige Danske Videnskabernes Selskabs Skrifter, Naturvidenskabelig og Mathematisk Afdeling*, *9 Raekke*, **2**, 231–326.

BOLZ, H. 1971a. Die Zlambach-Schichten (alpine Obertrias) unterbesonderer Berucksichtigung der Ostracoden, 1: Ostracoden der Zlambach-Schichten, besonders Bairdiidae. *Senckenbergiana Lethaea*, **52**, 129–283.

—— 1971b. Late Triassic Bairdiidae and Healdiidae, 717–745. *In* OERTLI, H. J. (ed.). Colloque sur la Paleoecologie des Ostracodes, Pau, July 20–27, 1970. *Bulletin du Centre de Recherches de Pau (Société Nationale des Petroles d'Aquitaine)*, **5**, 1–953.

BONAPARTE, C. L. 1832. *Iconografia della fauna italica, per le quatro classi degli animali vertebri.* Rome, 78 pp.

—— 1838. Synopsis vertebratorum systematis. *Nuovi Annali Scienze Naturali (Bologna)*, **2**, 105–133.

BOOMER, I. D. 1991a. Lower Jurassic ostracod biozonation of the Mochras Borehole. *Journal of Micropalaeontology*, **9**, 205–218.

—— 1991b. On *Ogmoconchella martini* (Anderson). *Stereo Atlas of Ostracod Shells*, **18**, 121–124.

—— 1991c. On *Ektyphocyhere cookiana* (Anderson). *Stereo Atlas of Ostracod Shells*, **18**, 113–116.

BORNEMANN, J. G. 1854. *Über die Liasformation in der Umgegend von Göttingen und ihre organischen Einschlüsse*. Berlin, 77 pp.

BOURNE, G. C. 1900. *An introduction to the study of the comparative anatomy of animals. Vol. 1. Animal organisation: the Protozoa and Coelenterata*. London, xvi+269 pp.

BRADY, G. S. and NORMAN, A. M. 1889. A monograph of the marine and freshwater Ostracoda of the north Atlantic and of northwestern Europe. Section 1: Podacopa. *Scientific Transactions of the Royal Dublin Society, Series 2*, **4**, 63–270.

—— and ROBERTSON, D. 1885. Genus *Darwinula*. *In* JONES, T. R. On the Ostracoda of the Purbeck formation; with notes on the Wealden species. *Quarterly Journal of the Geological Society, London*, **41**, 346.

BRADY, H. B. 1881. Notes on some of the reticularian Rhizopoda of the Challenger Expedition. Part 3. 1—Classification, 2—Further notes on new species, 3—Note on *Biloculina* mud. *Quarterly Journal of the Microscopical Society, New Series*, **21**, 31–71.

BRANDT, J. F. 1835. *Mammalium exoticorum novorum vel minus rite cognitorum Musei Academici Zoologici descriptiones et icones*. Vol. 1, 106 pp. Vol. 2, 85 pp., Petropoli.

BRANSON, C. C. 1942. *Parallelodon, Grammatodon* and *Beshausenia*. *Journal of Paleontology*, **16**, 247–249.

BRASIER, M. D. 1980. *Microfossils*. George Allen and Unwin Ltd, London, xii+193 pp.

BRAUN, C. F. W. 1843. Beiträge zur Urgeschichte der Pflanzen. 1–33. *In* MÜNSTER, G. G. zu (ed.). *Beiträge zur Petrefactenkunde, etc.*, 6. Bayreuth.

BRIGGS, D. E. G., CLARKSON, E. N. K. and ALDRIDGE, R. J. 1983. The conodont animal. *Lethaia*, **16**, 1–14.

BRODIE, P. B. 1845. *A history of the fossil insects in the secondary rocks of England*. John van Voorst, London, 130 pp.

—— 1872. On phosphatic and bone bed deposits in British strata, their economic uses and fossil contents. *Annual Report of the Warwickshire Natural History and Archaeological Society*, **36**, 53–67.

—— 1876. On a further extension of the Rhaetics (or Penarth Beds) in Warwickshire, Leicestershire, Nottinghamshire, Rutland, Lincolnshire, Yorkshire, and Cumberland, and on the occurrence of some supposed remains of *Labyrinthodon* and a new radiate therein. *Proceedings of the Warwickshire Naturalists' and Archaeologists' Field Club*, 1–11.

—— 1886. On the range, extent and fossils of the Rhaetic Formation in Warwickshire. *Proceedings of the Warwickshire Naturalists' and Archaeologists' Field Club*, 19–29.

BROMLEY, R. G. 1996. *Trace fossils*. 2nd edition. Chapman and Hall, London, xvi+361 pp.

BRONGNIART, A. 1849. *Tableau des genre de végétaux fossiles considérés sous le point de vue de leur classification botanique et de leur distribution géologique. Extrait du Dictionnaire d'Histoire naturelle*. Paris, 127 pp.

BRONN, H. G. 1848–49. *Handbuch einer Gesichte der natur. Dritter Band. 3 Teil*; *Orgánisches leben (Fortzetzung). Index Palaeontologicus.* 1 (1848), 'A-M', i–viii, lxxxiv+775 pp.; 2 (1848), 'N-Z', 776–1381; 3 (1849), 4 Teil, Vernunstleben, Stuttgart, 1106 pp.

—— 1860. *Klassen und Ordnungen des Thier-Reichs.* Vol. 1. *Amorphozoen.* Leipzig and Heidelberg, 434 pp.

BROOM, R. 1905. On the structure and affinities of the endothiodont reptiles. *Transactions of the South African Philosophical Society*, **15**, 259–282.

BROTZEN, F. 1948. The Swedish Paleocene and its foraminiferal fauna. *Sveriges Geologiska Undersökning, Series Ca, Avhandlingar och Uppsatser*, **42**, 1–140.

BROUGH, J. 1935. On the structure and relationships of the hybodont sharks. *Memoirs of the Manchester Literary and Philosophical Society*, **79**, 35–49.

BROWNE, M. 1890. A supplementary contribution of the history of the fossil fishes of the Rhaetics of Leicestershire. *Transactions of the Leicester Literary and Philosophical Society*, **2**, 178–179.

—— 1895. Excursion to East Leake and comparison of the section with Wigston. *Transactions of the Leicester Literary and Philosophical Society*, **4**, 52–53.

BRUGUIÈRE, J. G. 1791–92. *Tableau encyclopédique et méthodique des trois règnes de la nature contenant l'helminthologie, ou les vers infusoires, les vers intestins, les vers mollusques, etc.* ... livre 7, part 1. Panckouke, Paris, i+83 pp.

—— 1797. *Tableau encyclopédique et méthodique des trois règnes de la nature. Vol. 2, Vers, Coquilles, Mollusques et Polypiers.* Panckouke, Paris.

BUCKLAND, W. 1824. Notice on the *Megalosaurus*, or great fossil lizard of Stonesfield. *Transactions of the Geological Society, London*, **2**, 390–396.

—— 1829. On the discovery of coprolites, or fossil faeces, in the Lias at Lyme Regis, and in other formations. *Transactions of the Geological Society, London*, **3**, 223–236.

—— 1834. (On fossil faeces and fossil Sepia). *Proceedings of the Geological Society, London*, **1**, 142–143.

—— and CONYBEARE, W. D. 1824. Observations on the south-western coal district of England. *Transactions of the Geological Society, Series 2*, **1**, 210–316.

BUCKMAN, J. 1850. On some fossil plants from the Lower Lias. *Quarterly Journal of the Geological Society, London*, **6**, 413–418.

BURKENROAD, M. D. 1963. The evolution of the Eucarida (Crustacea, Eumalacostraca) in relation to the fossil record. *Tulane Studies in Geology*, **2**, 3–16.

BURMEISTER, H. 1834. *Beiträge zur Naturgeschichte des Rankenfüsser (Cirripedia).* Berlin, viii +60 pp.

—— 1836–37. *Handbuch der Naturgeschichte, etc.* Abt. I. *Mineralogie und Botanik.* Abt. 2. *Zoologie.* Berlin, viii, xxvi, xii+858 pp.

CALLAWAY, C. 1901. The pre-Rhaetic denudation of the Bristol area. *Proceedings of the Cotteswold Naturalists' Field Club*, **14**, 47–58.

CALMAN W. T. 1904. On the classification of the Crustacea Malacostraca. *Annals and Magazine of Natural History, Series 7*, **13**, 144–158.

CARTER, J. G. 1990. Evolutionary significance of shell microstructure in the Palaeotaxodonta, Pteriomorphia and Isofilibranchia (Bivalvia: Mollusca). 135–296. *In* CARTER, J. G. (ed.). *Skeletal biomineralization: patterns, processes and evolutionary trends.* Vol. 1. Van Nostrand Reinhold, New York, vii+832 pp.

CAUMONT, A. de 1825. Sur quelques terrains de la Normandie occidentale, lu à la Société Linnéenne du Calvados, dans les séances du 8 novembre 1824, du 3 janvier et du 7 février 1825. *Mémoires de la Société Linnéenne du Calvados*, **2**, 447–597.

CAVERS, F. 1910. The inter-relationships of the Bryophyta. I. Sphaerocarpales. *New Phytologist*, **10**, 81–92.

CHAPMAN, F. 1894. On Rhaetic foraminifera from Wedmore, in Somerset. *Annals and Magazine of Natural History, Series 6*, **16**, 305–329.

CHAPUIS, F. and DEWALQUE, G. 1851–3. Description des fossiles des terrains secondaires de la Province de Luxembourg. *Mémoire Couronné et Mémoires des savants étrangers, l'Académie Royale de Belgique, Bruxelles*, **25**, 1–303.

CHARIG, A. J., ATTRIDGE, J. and CROMPTON, A. W. 1965. On the origin of the sauropods and the classification of the Saurischia. *Proceedings of the Linnean Society*, **176**, 197–221.

CHAVAN, A. 1938. Essai critique de classification des lucines. *Journal de Conchyliologie*, **82**, 59–99, 105–130, 215–243.

—— 1954. Les *Pleurophorus* et genres voisins. *Cahiers Géologiques de Seyssel (Ain)*, **22**, 1–200.

—— 1969. Superfamilies Lucinacea, Carditacea, Crassatellacea. N491–N518, N543–N561, N562–N582. *In* MOORE, R. C. (ed.). *Treatise on invertebrate paleontology. Part N. Mollusca 6, Bivalvia (2)*. Geological Society of America and University of Kansas Press, Boulder, Colorado, and Lawrence, Kansas, ii+N491–N952.

CLAYPOOL, G. E. and KAPLAN, I. R. 1974. The origin and distribution of methane in marine sediments. 99–139. *In* KAPLAN, I. R. (ed.). *Natural gases in marine sediments*. Plenum Press.

CLEMENS, W. A. 1980. Rhaeto-Liassic mammals from Switzerland and West Germany. *Zitteliana*, **5**, 51–92.

—— and KIELAN-JAWOROWSKA, Z. 1979. Multituberculata. 99–149. *In* LILLEGRAVEN, J. A., KIELAN-JAWOROWSKA, Z. and CLEMENS, W. A. (eds). *Mesozoic mammals: the first two-thirds of mammalian history*. University of California Press, Berkeley, x+311 pp.

—— LILLEGRAVEN, J. A., LINDSAY, E. H. and SIMPSON, G. G. 1979. Where, when, and what – a survey of known Mesozoic mammal distribution. 7–58. *In* LILLEGRAVEN, J. A., KIELAN-JAWOROWSKA, Z. and CLEMENS, W. A. (eds). *Mesozoic mammals: the first two-thirds of mammalian history*. University of California Press, Berkeley, x+311 pp.

CLOUTIER. R. 1993. Interrelationships of Palaeozoic actinistians: patterns and trends. 379–428. *In* CHANG MEE-MANN, LIU YU-HAI and ZHANG GUO-RUI (eds). *Early vertebrates and related problems of evolutionary biology*. Science Press, Beijing, v +514 pp.

COMPAGNO, L. J. V. 1973. Interrelationships of living elasmobranchs. *Zoological Journal of the Linnean Society*, **53**, 15–61.

—— 1977. Phyletic relationships of living sharks and rays. 303–322. *In* Recent advances in the biology of sharks. *American Zoologist*, **17**, 287–515.

CONRAD, T. A. 1867. Descriptions of new genera and species of fossil shells. *American Journal of Conchology*, **3**, 8–16.

CONYBEARE, W. D. 1822. Additional notices on the fossil genera *Ichthyosaurus* and *Plesiosaurus*. *Transactions of the Geological Society, London, Series 2*, **1**, 103–123.

COPE, E. D. 1869. Synopsis of the extinct Batrachia, Reptilia, and Aves of North America. *Transactions of the American Philosophical Society*, **14**, 1–252.

—— 1871. Contribution to the ichthyology of the Lesser Antilles. *Transactions of the American Philosophical Society*, **14**, 445–483.

—— 1876. On some extinct reptiles and Batrachia from the Judith River and Fox Hills beds of Montana. *Proceedings of the Academy of Natural Sciences, Philadelphia, Paleontological Bulletin*, **23**, 340–359.

COPE, J. C. W. 1991*a*. Discussion on correlation of the Triassic-Jurassic boundary in England and Austria. *Journal of the Geological Society, London*, **148**, 420–422.

—— 1991*b*. Reply to E. G. Poole: further discussion on correlation of the Triassic-Jurassic boundary in England and Austria. *Journal of the Geological Society, London*, **148**, 943–944.

COPESTAKE, P. 1981. Triassic. 78–80. *In* JENKINS, D. G. and MURRAY, J. W. (eds). *Stratigraphical atlas of fossil foraminifera*. 1st edition. Ellis Horwood Ltd, Chichester, for the British Micropalaeontological Society, 310 pp.

—— 1989. Triassic. 97–124. *In* JENKINS, D. G. and MURRAY, J. W. (eds). *Stratigraphical atlas of fossil foraminifera*. 2nd edition. Ellis Horwood Ltd, Chichester, for the British Micropalaeontological Society, 593 pp.

—— and JOHNSON, B. 1981. Jurassic: the Hettangian to Toarcian. 81–105. *In* JENKINS, D. G. and MURRAY, J. W. (eds). *Stratigraphical atlas of fossil foraminifera*. 1st edition. Ellis Horwood Ltd, Chichester, for the British Micropalaeontological Society, 310 pp.

COSSMANN, M. 1895. Contribution à la Paléontologie Française des Terrains Jurassiques, Opisthobranchiata. *Mémoires de la Société Géologique de France*, **14**, 1–167.

—— 1896. *Essais de Paléoconchologie Comparée, deuxième livraison*. Paris, 179 pp.

—— 1909. *Essais de Paléoconchologie Comparée, huitième livraison*. Paris, 248 pp.

—— 1912. *Essais de Paléoconchologie Comparée, neuvième livraison*. Paris, 215 pp.

—— 1918. *Essais de Paléoconchologie Comparée, onzième livraison*. Paris, 388 pp.

COX, L. R. 1946. *Tutcheria* and *Pseudopis*, new lamellibranch genera from the Lias. *Proceedings of the Malacological Society, London*, **27**, 34–48.

—— 1949. On the occurrence of the genus *Tutcheria* in the Upper Trias and Rhaetic. *Proceedings of the Malacological Society, London*, **27**, 247–248.

—— 1960. General characters of Gastropoda. I84–I169. *In* MOORE, R. C. (ed.). *Treatise on invertebrate paleontology. Part I. Mollusca 1: Gastropoda*. Geological Society of America and University of Kansas Press, Boulder, Colorado and Lawrence, Kansas, I84–I351.

—— 1961. New genera and subgenera of Mesozoic Bivalvia. *Palaeontology*, **4**, 592–598.

—— 1963. The Rhaetic-Hettangian bivalve genus *Pteromya* Moore. *Palaeontology*, **6**, 582–595.

—— 1964. Notes concerning the taxonomy and nomenclature of fossil Bivalvia (mainly Mesozoic). *Proceedings of the Malacological Society, London*, **36**, 39–48.

—— NEWELL, N. D., BOYD, D. W., BRANSON, C. C., CASEY, R., CHAVAN, A., COOGAN, A. H., DECHASEAUX, C., FLEMING, C. A., HAAS, F., HERTLEIN, L. G., KAUFFMAN, E. G., KEEN, A. M., LaROCQUE, A., MCALESTER, A. L., MOORE, R. C., NUTTALL, C. P., PERKINS, B. F., PURI, H. S., SMITH, L. A., SOOT-RYEN, T., STENZEL, H. B., TRUEMAN, E. R., TURNER, R. D. and WEIR, J. 1969. *In* MOORE, R. C. (ed.). *Treatise on invertebrate paleontology. Part N. Mollusca 6, Bivalvia.* 2 vols. Geological Society of America and University of Kansas Press, Boulder, Colorado, and Lawrence, Kansas, xxxviii+N1–N489, ii+N491–N952.

CRUICKSHANK, A. R. I. 1996. A *Pistosaurus*-like sauropterygian from the Rhaeto-Hettangian of England. *Mercian Geologist*, **14**, 12–13.

CUNY, G. 1998. Primitive neoselachian sharks: a survey. *Oryctos*, **1**, 3–21.

—— and BENTON, M. J. 1999. Early radiation of the neoselachian sharks in western Europe. *Geobios*, **32**, 193–204.

—— MAZIN, J.-M. and RAUSCHER, R. 1994. Saint-Germain-les-Arlay: un nouveau site Rhétien date par la palynologie et l'étude des vertébrés dans le Département du Jura (France). *Revue de Paléobiologie*, **14**, 35–48.

CURTIS, C. D. 1980. Diagenetic alteration in black shales. *Journal of the Geological Society, London*, **137**, 189–194.

—— 1987. Mineralogical consequences of organic matter degradation in sediments in organic/organic diagenesis. 108–123. *In* LEGGETT, J. K. and ZUFFA, G. G. (eds). *Marine clastic sedimentology: concepts and case studies: a volume in memory of C. Tarquin Teale*. Graham and Trotman, London, x+211 pp.

CURTIS, M. T. 1981. The Rhaetic-Carboniferous Limestone unconformity at Southfields Quarry, Chipping Sodbury, Avon. *Proceedings of the Bristol Naturalists' Society*, **40**, 30–35.

CUSHMAN, J. A. 1910. A monograph of the Foraminifera of the North Pacific Ocean. Part 1. Astorhizidae and Lituolidae. *Bulletin of the United States National Museum*, **71**, 1–134.

—— 1927. An outline of a re-classification of the Foraminifera. *Contributions to the Cushman Laboratory for Foraminiferal Research*, **3**, 1–105.

—— and OZAWA, Y. 1930. A monograph of the foraminiferal family Polymorphinidae, recent and fossil. *Proceedings of the United States National Museum*, **77**, 1–185.

DALL, W. H. 1871. Report on the Brachiopoda obtained by the United States Coast Survey Expedition in charge of L. F. de Pourtalès, with a revision of the Craniidae and Discinidae. *Bulletin of the Harvard University Museum of Comparitive Zoology*, **3**, 1–45.

—— 1889a. On the hinge of pelecypods and its development, with an attempt toward a better subdivision of the group. *American Journal of Science*, **38**, 445–462.

—— 1889b. Reports on the results of dredging. 29. Report on the Mollusca, Part 2: Gastropoda and Scaphopoda. *Bulletin of the Museum of Comparative Zoology, Harvard University*, **18**, 1–492.

—— 1896–1900. Pelecypoda. 346–429. *In* ZITTEL, K. A. von (ed.). *Textbook of Palaeontology* (translated and edited by EASTMAN, C. R.). Vol. 1. Macmillan and Co., New York and London.

DAMES, W. 1888. Die Ganoiden des deutschen Muschelkalkes. *Geologische und Palaeontologische Abhandlungen*, **4**, 133–179.

DARWIN, C. 1854. *A monograph on the subclass Cirripedia, with figures of all the species: the Balanidae (or sessile Cirripedes): the Verrucidae*, etc. Ray Society, London, viii+684 pp.

DAVIDSON, T. 1851–52. A monograph of the British fossil Brachiopoda. Vol. 1, part 3, Oolitic and Liasic. *Monograph of the Palaeontographical Society*, **4** (10), 1–64, pls 1–13; **6** (17), 65–100, pls 14–18.

—— 1855. Appendix and Index to Vol. 1 of a monograph of the British fossil Brachiopoda. *Monograph of the Palaeontographical Society*, **8** (26), 1–30, pl. A.

—— 1876/78. Supplement to the British fossil Brachiopoda. Vol. 4, pt 2, Jurassic-Triassic. *Monograph of the Palaeontographical Society*, **30** (135), 73–144, pls 9–16; **32** (146), 145–242, pls 17–29.

DAVIS, J. W. 1881. Notes on the fish remains of the Bone Bed at Aust near Bristol; with a description of some new genera and species. *Quarterly Journal of the Geological Society, London*, **37**, 414–426.

DAWKINS, W. B. 1864. On the Rhaetic Beds and White Lias of western and central Somerset; and on the discovery of a new fossil mammal in the Grey Marlstones beneath the Bone-bed. *Quarterly Journal of the Geological Society, London*, **20**, 396–412.

DAWSON, J. W. 1873. Impressions and footprints of aquatic animals and imitative markings on Carboniferous rocks. *American Journal of Science, Series 3*, **5**, 16–24.

DECHEN, H. von 1832. *H. T. de la Beche's 'Handbuch der Geologie, bearbeitet von H. von Dechen*. Berlin.

DELAGE, Y. and HÉROUARD, E. 1896. *Traité de Zoologie Concrète. Tome 1. La cellule et les Protozoaires*. Paris, 584 pp.

DEPÉRET, C. and MAZERAN, P. 1912. Les Estheria du Permien d'Autun. *Bullétin de la Société d'Histoire Naturelle d'Autun*, **25**, 167–173.

DESHAYES, G. P. 1831. *Description des Coquilles caractéristiques des Terrains*. Paris, vii+264 pp.

DESOR, P. J. E. 1855–1858. *Synopsis des Échinides fossiles*. Paris and Wiesbaden, lxviii+490 pp.

DIETRICH, W. O. 1926. Steinkorallen des Malms und Unterkreide im südlichen deutsch-Ostafrika. *Palaeontographica*, **7**, 40–102.

DITTMAR, A. von 1864. *Die Contorta-zone (Zone der Avicula contorta, Portl.) ihre Verbreitung und ihre organischen Einschlüsse*. Munich, 217 pp.

DONOVAN, D. T. 1958. Easter Field Meeting: the Lower and Middle Jurassic rocks of the Bristol District. *Proceedings of the Geologists' Association*, **69**, 130–140.

—— and KELLAWAY, G. A. 1984. Geology of the Bristol district: the Lower Jurassic rocks. *Memoir of the British Geological Survey*, 1–69.

—— CURTIS, M. T. and CURTIS, S. A. 1989. A psiloceratid ammonite from the supposed Triassic Penarth Group of Avon, England. *Palaeontology*, **32**, 231–235.

—— HORTON, A. and IVIMEY-COOK, H. C. 1979. The transgression of the Lower Lias over the northern flank of the London platform. *Journal of the Geological Society of London*, **136**, 165–173.

DOUVILLÉ, H. 1904. Les explorations de M. de Morgon en Perse. *Bulletin de la Société Géologique de France, 4ème Série*, **4**, 539–553.

—— 1913. Classification des lamellibranches. *Bulletin de la Société Géologique de France, 4ème Série*, **12**, 419–467.

DUFF, P. 1842. *Sketch of the geology of Moray*. Forsyth and Young, Elgin, vi+72 pp.

DUFFIN, C. J. 1978a. *Tropifer laevis* Gould (Coleiidae: Crustacea) and a new crustacean from the Rhaetian bone bed at Aust Cliff, Avon. *Zoological Journal of the Linnean Society*, **64**, 177–185.

—— 1978b. The Bath geological collections. f. The importance of certain vertebrate fossils collected by Charles Moore: an attempt at scientific perspective. *Newsletter of the Geological Curators Group*, **2**, 59–67.

—— 1979. Coprolites: a brief review with reference to specimens from the Rhaetic bone-beds of England and South Wales. *Mercian Geologist*, **7**, 191–204.

—— 1980. The Upper Triassic section at Chilcompton, Somerset, with notes on the Rhaetic of the Mendips in general. *Mercian Geologist*, **7**, 251–267.

—— 1982a. A palaeospinacid shark from the Upper Triassic of south-west England. *Zoological Journal of the Linnean Society*, **74**, 1–7.

—— 1982b. Teeth of a new selachian from the Upper Triassic of England. *Neues Jahrbuch für Geologie und Paläontologie, Monatshefte*, **1982**, 156–166.

—— 1985. Revision of the hybodont selachian genus *Lissodus* Brough (1935). *Palaeontographica, Abteilung A*, **188**, 105–152.

—— 1993a. Late Triassic sharks teeth (Chondricthyes, Elasmobranchii) from Saint-Nicolas de Port (north-east France). *Professional Paper of the Belgian Geological Survey*, **264**, 7–32.

—— 1993b. Mesozoic chondrichthyan faunas. 1. Middle Norian (Upper Triassic) of Luxembourg. *Palaeontographica, Abteilung A*, **229**, 15–36.

—— 1993c. Reworked Courceyan (Early Carboniferous) chondrichthyan remains from British Triassic cave deposits. *In* TURNER, S. (ed.). *IGCP328: SDS The Gross Symposium, Göttingen 31 July–7 Aug. 1993*. Abstracts, 2 pp.

—— 1993d. Teeth of *Hybodus* (Selachii) from the Early Jurassic of Lyme Regis, Dorset (southern England): preliminary note. *Professional Paper of the Belgian Geological Survey*, **264**, 45–52.

—— 1994. Myriacanthid holocephalans (Chondrichyes) from the British Late Triassic. *Neues Jahrbuch für Geologie und Paläontologie, Abhandlungen*, **192**, 1–16.

—— 1998a. New fossil shark remains from the British Rhaetian (latest Triassic). 1. The earliest basking shark. *Neues Jahrbuch für Geologie und Paläontologie, Monatshefte*, **1998**, 157–181.

—— 1998b. New fossil shark remains from the British Rhaetian (latest Triassic). 2. Hybodonts and palaeospinacids. *Neues Jahrbuch für Geologie und Paläontologie, Monatshefte*, **1998**, 240–256.

—— and DELSATE, D. 1993. The age of the Upper Triassic vertebrate fauna from Attert (Luxembourg, Belgium). *Professional Paper of the Belgian Geological Survey*, **264**, 33–44.

—— and FURRER, H. 1981. Myriacanthid holocephalan remains from the Rhaetian (Upper Triassic) and Hettangian (Lower Jurassic) of Graubunden (Switzerland). *Eclogae Geologicae Helvetiae*, **74**, 803–829.

—— and WARD, D. J. 1993. The Early Jurassic palaeospinacid sharks of Lyme Regis, southern England. *Professional Paper of the Belgian Geological Survey*, **264**, 53–102.

—— COUPATEZ, P., LEPAGE, J.-C. and WOUTERS, G. 1983. Rhaetian (Upper Triassic) marine faunas from 'Le golfe du Luxembourg' in Belgium (preliminary note). *Bulletin de la Société Belge de Geologie*, **92**, 311–315.

DUMÉRIL, A. M. C. 1806. *Zoologie analytique ou méthode naturelle de classification des animaux*. Allais, Paris, xxiv+344 pp.

DUMORTIER, E. 1864. *Études paléontologiques sur les Dépots Jurassiques du Bassin du Rhône. Part 1. Infra Lias*. Paris, 187 pp.

DUNCAN, P. M. 1867–68. A monograph of the British fossil corals, Supplement. Part 4. Liassic. *Monograph of the Palaeontographical Society*, **20** (85), i–iii, 1–44, pls 1–11; **21** (90), 45–73, pls 12–17.

—— 1889. A revision of the genera and great groups of the Echinoidea. *Zoological Journal of the Linnean Society*, **23**, 1–311.

DUNKER, W. 1844. Vorläufige, Diagnosen mehrerer neuer Conchylien aus der Nord-Deutschen Liasbildung die nächstens aus führlicher beschrieben und abgebildet erscheinen werden. *Zeitschrift für Malakozoologie*, **1**, 186–188.

—— 1846–47. Über die in dem Lias bei Halberstadt vorkommenden Versteinerungen. *Palaeontographica*, **1**, (1846), 34–41; (1847), 107–112, 128–133 (Über einige neue Versteinerungen aus verschiedenen Gebirgsformationen).

EASTWOOD, T., WHITEHEAD, T. H. and ROBERTSON, T. 1925. The geology of the country around Birmingham. *Memoirs of the Geological Survey, England and Wales*, 1–152.

EDMUNDS, E. A. and WILLIAMS, B. J. 1985. Geology of the country around Taunton and the Quantock Hills. *Memoir of the British Geological Survey*, i–xii+1–92.

—— POOLE, E. G. and WILSON, V. 1965. Geology of the country around Banbury and Edge Hill. *Memoirs of the Geological Survey of Great Britain*, 1–137.

EGERTON, P. 1854. On some new genera and species of fossil fishes. *Annals and Magazine of Natural History*, **13**, 433–436.

EHRENBERG, C. G. 1834. *Die Corallenthiere des Rothen Meeres physiologisch untersucht und systematisch verzeichnet*. Berlin, 156pp. (reprinted from *Abhandlungen der Koniglichen Akademie der Wissenschaften in Berlin*, 1832 (1834), with the addition of a title-page).

—— 1838. Über dem blossen Auge unsichtbare Kalkthierchen und Kieselthierchen als Hauptbestandtheile der Kreidegebirge. *Bericht über die zur Bekanntmachung geeigneten Verhandlungen der Koniglich Preussischen Akademie der Wissenschaften zu Berlin*, **3**, 192–200.

EHRENBERG, K. 1944. Ergänzende Bemerkungen zu den seinerzeit aus dem Miozän von Burgschleinitz beschriebenen Gangkernen und Bauten dekapoder Krebse. *Paläontologische Zeitschrift*, **23**, 354–359.

EICHWALD, C. E. von 1829–31. *Zoologia specialis quam expositis Animalibus tum vivis, tum fossilibus potissimum Rossiæ in universum, et Poloniæ in specie*, etc. Vol. 2, 21. Viln.

EMMRICH, A. 1853. Geognostiche Beobachtungen aus den östlichen bayerischen und den angränzenden österrchischen Alpen. II, Aus dem Gebiete des Alpenkalkes. *Jahrbuch der Kaiserlich-Königlichen geologischen Reichsanstalt Wien*, **4**, 326–394.

ENGLER, A. 1892. *Syllabus der Vorlesungen über spezielle und medizinisch-pharmaceutische Botanik. Eine verber sicht über das gesamie Pflanzensystem mit Berücksichtigung der Medicinal-und Nutzpflanzen*. Bornträger, Berlin, xxiii +184 pp.

ÉTALLON, A. 1862. Études paléontologiques sur les terrains Jurassiques du Haut-Jura. Monographie de l'étage Corallien. 1. Vertébrés, Articulés, Mollusques. *Mémoires de la Société d'Emulation du Département du Doubs, Paris, 3ème Série*, **6**, 1–250.

ETHERIDGE, R. 1865. On the Rhaetic or *Avicula contorta* Beds at Garden Cliff, Westbury-on-Severn, Gloucestershire. *Proceedings of the Cotteswold Naturalists' Field Club*, **3**, 218–234.

—— 1872. On the physical structure and organic remains of the Penarth (Rhaetic) Beds of Penarth and Lavernock; also with description of the Westbury-on-Severn section. *Report and Transactions of the Cardiff Naturalists' Society*, **3**, 39–64.

—— 1881. On the presence of the scattered skeletal remains of Holothuroidea in the Carboniferous Limestone Series of Scotland. *Proceedings of the Royal Physical Society, Edinburgh*, **6**, 183–198.

FABRICIUS, J. C. 1793. Om Skrifter i Insektlaren. *Naturhistorie-Selskab. Skrivter*, **3**, 145–156.

FÅHRÆUS, L. E. and RYLEY, C. C. 1989. Multielement species of *Misikella* Kozur and Mock, 1974 and *Axiothea* n. gen. (Conodonta) from the Mamonia Complex (Upper Triassic), Cyprus. *Canadian Journal of Earth Sciences*, **26**, 1255–1263.

FALCON, N. L. and KENT, P. E. 1960. Geological results of petroleum exploration in Britain, 1945–1957. *Memoirs of the Geological Survey of Great Britain*, **2**, 1–56.

FELL, H. B. 1966. Diadematacea. U340–U366. *In* MOORE, R. C. (ed.). *Treatise on invertebrate paleontology. Part U. Echinodermata 3 (1)*. Geological Society of America and University of Kansas Press, Boulder, Colorado and Lawrence, Kansas, xxx+366 pp.

FÉRUSSAC, A. E. de 1821–22. *Tableaux systématiques des animaux mollusques classés en familles naturelles, dans lesquels on a établi la concordance de tous les systèmes; suivis d'un Prodrome général pour tous les mollusques terrestres ou fluviatiles, vivants ou fossiles.* A. Bertrand, Paris and London, xlviii+111 pp.

FISCHER, P. H. 1880–87. *Manuel de conchyliologie et de paléontologie conchyliologique ou histoire naturelle des mollusques vivants et fossiles suivi d'un appendice sur les brachiopodes par D. P. Oehlert.* F. Savy, Paris, xxiv +1369 pp.

FLEMING, J. 1828. *A history of British animals, exhibiting the descriptive characters and systematical arrangement of the genera and species of quadrupeds, birds, reptiles, fishes, Mollusca and Radiata of the United Kingdom; including the indigenous, extirpated, and extinct kinds; together with periodical and occasional visitants.* Bell and Bradfute, Edinburgh, xxiii+565 pp.

FÖRSTER, R and CRANE, M. D. 1984. A new species of the penaeid shrimp *Aeger* Münster (Crustacea, Decapoda) from the Upper Triassic of Somerset, England. *Neues Jahrbuch für Geologie und Paläontologie, Monatshefte*, **1984**, 455–462.

FOWLER, A. and ROBBIE, J. A. 1961. Geology of the country around Dungannon. *Memoirs of the Geological Survey of Northern Ireland*, i–xii+1–274.

FRANKE, A. 1936. Die Foraminiferan des deutschen Lias. *Abhandlungen der Preussischen geologischen Landesanstalt, Neue Folge*, **169**, 1–138.

FRASER, N. C. 1985. Vertebrate faunas from the Mesozoic fissure deposits of south west Britain. *Modern Geology*, **9**, 273–300.

—— 1986. Terrestrial vertebrates at the Triassic-Jurassic boundary in south west Britain. *Modern Geology*, **10**, 147–157.

—— and WALKDEN, G. M. 1983. The ecology of a Late Triassic reptile assemblage from Gloucestershire, England. *Palaeogeography, Palaeoclimatology, Palaeoecology*, **42**, 341–365.

FRITSCH, A. 1907. *Miscellanea palaeontologica, 1. Palaeozoica.* Prague, 23 pp.

FRIZZELL, D. L. and EXLINE, H. 1956. Monograph of fossil Holothurian sclerites. *Bulletin of the Missouri University School of Mines and Metallurgy, Technical Series*, **89**, 1–204.

FUCHS, W. 1970. Eine Alpine, tiefliassische Foraminiferenfauna von Hemstein in Niederösterreich. *Verhandlungen der Geologischen Bundesanstalt, Wien*, **1**, 66–145.

GABBOTT, S. E., ALDRIDGE, R. J. and THERON, J. N. 1995. A giant conodont with preserved muscle tissue from the Upper Ordovician of South Africa. *Nature*, **374**, 800–803.

GALLOIS, R. W. 1965. *British Regional Geology: the Wealden district.* 4th edition. HMSO, for Institute of Geological Sciences, London, i–xii+1–101.

—— and WORSSAM, B. C. 1993. The geology of the country around Horsham. *Memoir of the British Geological Survey*, i–viii+1–130.

GALTON, P. M. 1985. Notes on the Melanorosauridae, a family of large prosauropod dinosaurs (Saurischia: Sauropodomorpha). *Geobios*, **18**, 671–676.

GARDINER, B. G. 1960. A revision of certain actinopterygian and coelacanth fishes, chiefly from the Lower Lias. *Bulletin of the British Museum (Natural History), Geology*, **4**, 241–384.

—— 1961. New Rhaetic and Liassic beetles. *Palaeontology*, **4**, 87–89.

GARDNER, J. S. 1886. On Mesozoic Angiosperms. *Geological Magazine*, dec. 3, **3**, 193–204.

GAUNT, G. D., IVIMEY-COOK, H. C., PENN, I. E. and COX, B. M. 1980. Mesozoic rocks proved by IGS boreholes in the Humber and Acklam areas. *Report of the Institute of Geological Sciences of Great Britain*, **79/13**, i–ii+1–34.

GEER, C. de 1752–78. *Mémoires pour servir a l'histoire des Insectes.* 7 vols. Stockholm.

GEIGER, M. E. and HOPPING, C. A. 1968. Triassic stratigraphy of the southern North Sea Basin. *Philosophical Transactions of the Royal Society of London, Series B*, **254**, 1–36.

GEOFFROY, S.-H. E.-F. 1833. Divers mémoires sur les grands sauriens trouvés à l'état fossile vers les confins maritimes de la Basse-Normandie, attribués d'abord au crocodile, puis déterminés sous les noms de *Teleosaurus* et *Steneosaurus*. *Mémoires de l'Académie des Sciences, Paris*, **12**, 3–138.

GERSTAECKER, K. E. A. 1866. Gliedenfüssler (Arthropoda). *In* BRONN, H. G. *Klassen und Ordnungen des Tierreichs.* 1st edition. Vol. 5, part 1. Akademie Verlag, Leipzig, 1029 pp.

GIEBEL, C. G. 1856. *Fauna der Vorwelt mit steter Berücksichtigung der lebenden Thiere. Band 2 Abt. 1. Die Insecten und Spinnen.* Leipzig, xviii+511 pp.

GILL, T. 1862. Analytical synopsis of the Order Squali; and revision of the nomenclature of the genera. Squalorum generum novorum descriptiones diagnosticae. *Annals of the Lyceum of Natural History of New York*, **7**, 367–413.

—— 1872. Arrangement of the families of fishes, or classes Pisces, Marsipo-branchii, and Leptocardii. *Smithsonian Miscellaneous Collections*, **247**, i–xlvi+1–49.

GILLILAND, P. M. 1992. Holothurians in the Blue Lias of southern Britain. *Palaeontology*, **35**, 159–210.

GLÜCKMAN, L. S. 1964. [*Sharks of the Paleogene and their stratigraphic significance.*] Nauka Press, Moscow and Leningrad, 229 pp. [In Russian].

GOETEL, W. 1916. Die Rhätische Stufe und der unterste Lias der subtatrischen Zone in der Tatra. *Bulletin International de l'Académie des Sciences de Cracovie, Classe des Sciences Mathématiques et Naturelles, Série A, Sciences Mathématiques*, 1–222.

GOLDFUSS, G. A. 1826–44. *Petrefacta Germaniae tam ea, quae in Museo Universitatis Regiae Borussicae Fredericiae Wilhelmiae Rhenanae servantur quam alia quaecunque in muscis Hoeninghusiano Muensteriano aliisque extant, iconibus et descriptionibus illustrata. Abbildungen und Beschreibungen der Petrefacten Deutschlands und der angränzenden Länder, unter Mitwirkung des Herrn Grafen Georg zu Münster herausgegeben.* I–III, text and Atlas. 1–68 (1833); 69–140 (1836); 141–224 (1837); 225–312 (1840). Arnz and Co., Düsseldorf.

GOLIKOV, A. N. and STAROBOGATOV, Y. I. 1975. Systematics of prosobranch gastropods. *Malacologia*, **15**, 185–232.

GOODRICH, E. S. 1909. Vertebrata, Craniata, fasc. I: 'Cyclostomes and fishes'. *In* LANKESTER, E. R. (ed.). *A treatise on zoology*. Part 10. A. and C. Black, London, 534 pp.

GORJANSKY, Y. YU. and POPOV, L. E. 1985. [The morphology, systematic position and origin of inarticulate brachiopods with carbonate shells.] *Paleontologicheskii Zhurnal*, **1985**, 3–13. [In Russian].

GOULD, C. 1857. On a new fossil crustacean (*Tropifer laevis* C. Gould) from the Lias bone bed. *Quarterly Journal of the Geological Society, London*, **13**, 360–363.

GRAMM, M. N. 1975. Marine Triassic Cytheracea (Ostracoda) from South Primorye (Soviet far east). *Palaeontographica, Abteilung A, Paläozoologie, Stratigraphie*, **151**, 102–110.

GRAY, J. E. 1824. A supplement to the appendix – shells. *In* PARRY, Sir W. E. *Journal of a voyage for the discovery of a north west passage performed in 1818–20, in HMS Hecla and Griper; with an appendix.* John Murray, London, clxxxi+ccxv.

—— 1840*a*. A synopsis of the genera and species of the class Hypostoma (Asterias Linn.). *Annals and Magazine of Natural History*, **6**, 175–184, 275–290.

—— 1840*b*. *Synopsis of the contents of the British Museum.* 42nd edition. London, 370 pp.

—— 1847. A list of the genera of Recent Mollusca, their synonyms and types. *Proceedings of the Zoological Society, London*, **15**, 129–219.

—— 1850. *Catalogue of the Mollusca* [*including fossil forms*] *in the collection of the British Museum. Part 2*: Pteropoda. London, iv+45 pp.

GRIFFITH, H. E. and WILSON, H. E. 1982. Geology of the country around Carrickfergus and Bangor. *Geological Survey of Northern Ireland*, i–x+1–118.

GRUBE, A. E. 1850–51. *Die Familien der Anneliden mit Angabe ihrer Gattungen und Arten.* Berlin, 164 pp.

GRÜNDEL, J. 1964. Zur Gattung *Healdia* (Ostracoda) und zu einigen verwandten Formen aus dem unteren Jura. *Geologie,* **13,** 456–477.

GUINOT, D. 1977. Propositions pour une nouvelle classification des crustacés décapodes brachyoures. Heterotremata et Thoracotremata Guinot, 1977. *Comptes réndus hebdomadaires des Séances de l'Académie des sciences, Paris, Série D, Sciences Naturelles,* **285,** 1049–1052.

HAAN, W. de 1833–50. Crustacea. *In* SIEBOLD, P. F. de *Fauna Japonica sive descriptio animalium, quae in itinere per Japoniam, jussu et auspiciis superiorum, qui summum in India Batava Imperium tenent, suscepto, annis 1823–1830 collegit, notis, observationibus et adumbrationibus illustravit.* A. Arnz (Lugdunum Batavorum), ix–xvi, i–xxxi, vii–xvii+243 pp.

HAECKEL, E. 1866. *Generalle Morphologie der Organismen. Allgemeine Grundzüge der organischen Formen-Wissenschaft, mechanisch begründet durch die von Charles Darwin reformierte Deszendenz-Theorie. II. Allgemeine Entwicklungsgeschichte der Organismen. Kritische Gründzuge der mechanischen Wissenschaft von den entstehenden Formen der Organismen, begründet durch die Deszendenz-Theorie.* Berlin, clx+462 pp.

—— 1894. *Systematische Phylogenie. Entwurf eines natürlichen Systems der Organismen auf Grund ihrer Stammesgeschichte. Theil 1. Systematische Phylogenie der Protisten und Pflanzen.* Georg Reimer, Berlin, xv+400 pp.

HAEUSLER, R. 1881. *Untersuchungen ueber die microscopischen Structurverhaeltnisse der Aargauer Jurakalke mit besonderer Beruecksichtigung ihrer Foraminiferenfauna: Inaugural-Dissertation vorgelegt der hohen philesophischen Facultaet der Universitaet Zurich von Rudolf Haeusler.* Fisch, Wild and Co., Brugg, 47 pp.

HALDEMAN, S. 1840. *Supplement to number one of 'A Monograph of the Limniades, and other freshwater univalve shells of North America',* containing descriptions of apparently new animals in different classes, and the names and characters of the subgenera in Paludina and Anculosa. Philadelphia, 3 pp.

HALL, J. 1847–52. *Palaeontology of New York.* Vol. 1, 338 pp. (1847); vol. 2, 362 pp. (1852). Von Benthuysen, Albany, New York.

HALLAM, A. 1960. The White Lias of the Devon coast. *Proceedings of the Geologists' Association,* **71,** 47–60.

—— 1990. Correlation of the Triassic-Jurassic boundary in England and Austria. *Journal of the Geological Society, London,* **147,** 421–424.

—— 1991. Discussion of the Triassic-Jurassic boundary in England and Austria. *Journal of the Geological Society, London,* **148,** 420–422.

—— and EL SHAARAWY, Z. 1982. Salinity reduction of the end-Triassic sea from the Alpine region into northwestern Europe. *Lethaia,* **15,** 169–178.

HALSTEAD, L. B. and NICOLL, P. G. 1971. Fossilised caves of Mendip. *Studies in Speleology,* **2,** 93–102.

HAMBLIN, R. J. O., CROSBY, A., BALSON, P. S., JONES, S. M., CHADWICK, R. A., PENN, I. E. and ARTHUR, M. J. 1992. *United Kingdom offshore regional report: the geology of the English Channel.* HMSO, for the British Geological Survey, London, x+106 pp.

HAMILTON, D. 1961. Algal growths in the Rhaetic Cotham Marble of southern England. *Palaeontology*, **4**, 324–333.

—— 1962. Some notes on the Rhaetic sediments of the Filton by-pass substitute, near Bristol. *Proceedings of the Bristol Naturalists' Society*, **30**, 279–285.

—— 1977. Aust Cliff. 110–118. *In* SAVAGE, R. J. G. (ed.). *Geological excursions in the Bristol district*. University of Bristol Printing Unit, Bristol, xv+196 pp.

HAMILTON, G. B. 1982. Triassic and Jurassic calcareous nannofossils. 17–39. *In* LORD, A. R. (ed.). *A stratigraphical index of calcareous nannofossils*. Ellis Horwood Ltd, Chichester, 192 pp.

HANDLIRSCH, A. 1906–08. *Die fossilen Insekten und die Phylogenie der rezenten Formen*. Leipzig, ix+1430 pp.

—— 1939. Neue Untersuchungen über die Fossilen Insekten. 2. *Annalen der naturhistorisches Museum, Wien*, **49**, 1–240.

HÄNTZSCHEL, W. 1975. Trace fossils and problematica. W1–W269. *In* MOORE, R. C. and TEICHERT, C. (eds). *Treatise on invertebrate paleontology. Part W, Miscellanea, supplement 1*. Geological Society of America and University of Kansas Press, Boulder, Colorado and Lawrence, Kansas.

HARLAND, W. B., ARMSTRONG, R. L., COX, A. V., CRAIG, L. E., SMITH, A. G. and SMITH, D. G. 1990. *A Geologic Time Scale 1989*. Cambridge University Press, Cambridge, 263 pp.

HARLTON, B. H. 1933. Micropaleontology of the Pennsylvanian Johns Valley Shale of the Quachita Mountains, Oklahoma, and its relationship to the Mississippian Caney Shale. *Journal of Paleontology*, **7**, 3–29.

HARRIS, T. M. 1931. Rhaetic floras. *Biological Reviews of the Cambridge Philosophical Society*, **6**, 133–162.

—— 1937. *Naiadita*, a strange fossil bryophyte. *Report of the British Association for the Advancement of Science*, **1937**, 429.

—— 1938. *The British Rhaetic (Triassic) flora*. British Museum (Natural History), London, 84 pp.

—— 1939. *Naiadita*, a fossil bryophyte with reproductive organs. *Annals of Bryology*, **12**, 57–70.

—— 1957. A Liasso-Rhaetic flora in South Wales. *Proceedings of the Royal Society, Series B*, **147**, 289–308.

—— and JACKSON, J. S. 1961. On *Otozamites bechei* Brongniart from the Irish Rhaetic. *Proceedings of the Royal Irish Academy*, **61B**, 339–344.

HARRISON, W. J. 1876. On the occurrence of the Rhaetic Beds in Leicestershire. *Quarterly Journal of the Geological Society, London*, **32**, 212–218.

—— 1880. A starfish bed in the Rhaetic Formation. *Science Gossip*, 56.

HATSCHEK, B. 1888–91. *Lehrbuch der Zoologie, eine morphologische Übersicht des Theirreiches zur Einfürung in das Studium dieser Wissenschaft, Lief 1–3*. Gustav Fischer, Jena, iv +432 pp.

—— 1893. System der Anneliden, ein vorlaeufiger Bericht. *Lotos, Prag* **13**, 123–126.

HAUER, F. R. von 1853. Ueber die Gliederung der Trias, Lias, und Jura gebilde in den nordöstlichen Alpen. *Jahrbuch der Kaiserlich-Königlichen geologischen Reichsanstalt, Wien*, **4**, 715–784.

274   *Fossils of the Rhaetian Penarth Group*

HAY, O. P. 1902. Bibliography and catalogue of the fossil Vertebrata of North America. *Bulletin of the United States Geological Survey*, **179**, 1–868.

HEDLEY, C. 1918. A check-list of the fauna of New South Wales, Part 1. *Transactions and Proceedings of the Royal Society of New South Wales*, **51**, M1–M120.

HEMSLEY, A. R. 1989. The ultrastructure of the spore wall of the Triassic bryophyte *Naiadita lanceolata*. *Review of Palaeobotany and Palynology*, **61**, 89–99.

HERMAN, J. 1975. Zwei neue Haifischzähne aus der Kreide von Misburg bei Hannover (höheres Campan). *Bericht der Naturhistorische Gesellschaft Hannover*, **119**, 295–302.

HERTWIG, C. W. T. R. 1895. *Lehrbuch der Zoologie*. 3te Auflage, Jena, xii+599 pp.

HESS, H. 1965. Trias-Ophiuren aus Deutschland, England, Italien und Spanien. *Mitteilungen der Bayerischen Staatssammlung für Paläontologie und Historische Geologie*, **5**, 151–177.

HIRSCHMANN, C. 1959. Über Conodonten aus dem Oberen Muschelkalk des Thüringer Beckens. *Freiberger Forschungshefte*, **76**, 33–86.

HODGES, P. 1991. The relationship of the Mesozoic bivalve *Atreta* to the Dimyidae. *Palaeontology*, **34**, 963–970.

—— 1994. The base of the Jurassic System: new data on the first appearance of *Psiloceras planorbis* in southwest Britain. *Geological Magazine*, **131**, 841–844.

HORTON, A., POOLE, E. G., WILLIAMS, B. J., ILLING, V. C. and HOBSON, G. D. 1987. Geology of the country around Chipping Norton. Memoir for 1:50 000 geological sheet 218, New Series (England and Wales). *Memoir of the British Geological Survey*, i–x+1–169.

HORWOOD, A. R. 1912. On *Archarenicola rhaetica* sp. nov. *Geological Magazine*, dec. 5, **9**, 395–399.

—— 1913/1916. The Upper Trias in Leicestershire. *Geological Magazine*, dec. 5, **10**; 21–32, 73–86, 109–21, 205–15; dec. 6, **3**, 360–371, 411–420, 456–462.

HUDLESTON, W. H. and WILSON, E. 1892. *A catalogue of the British Jurassic Gastropoda*. Dulau and Co., London, 147 pp.

HUENE, E. von 1933. Zur Kenntnis der Württembergischen Rhätbonebeds mit Zahnfunden neuer Säuger und säugerzähnlicher Reptilien. *Jahreshefte des Vereins für vaterländische Naturkunde in Württemberg*, **1933**, 65–128.

—— 1935. Ein Rhynchocephale aus dem Rhät (*Pachystropheus* n. g.). *Neues Jahrbuch für Mineralogie, Geologie und Paläontologie*, **74**, 441–447.

HUENE, F. R. F. von 1902. Uebersicht über die Reptilien der Trias. *Geologische und paläontologische Abhandlungen*, **6**, 1–84.

—— 1908. On phytosaurian remains from the Magnesian Conglomerate of Bristol (*Rileya platyodon*). *Annals and Magazine of Natural History*, **8**, 228–230.

—— 1928. Kurze Übersicht über die Saurischia und ihre natürlichen Zusammenhänge. *Palaeontologische Zeitschrift*, **11**, 269–273.

—— 1932. Die fossile Reptil-Ordnung Saurischia, ihre Entwicklung und Geschichte. *Monographien zur Geologie und Palaeontologie*, **4**, 1–361.

HUNT, D. and TUCKER, M. E. 1992. Stranded parasequences and the forced regressive wedge systems tract: deposition during base level fall. *Sedimentary Geology*, **81**, 1–9.

HUXLEY, T. H. 1869. Triassic Dinosauria. *Nature*, **1**, 23–24.

—— 1875. On *Stagonolepis robertsoni*, and on the evolution of the Crocodilia. *Quarterly Journal of the Geological Society, London*, **31**, 423–438.

—— 1880. On the application of the laws of evolution to the arrangement of the Vertebrata and more particularly of the Mammalia. *Proceedings of the Zoological Society of London*, **1880**, 649–662.

I.C.Z.N. 1955. Validation, under the Plenary Powers, of the generic name *Modiolus* Lamarck, 1799 (Class Pelecypoda). *In* HEMMING, F. (ed.). Opinion 325. *Opinions and Declarations rendered by the International Commission on Zoological Nomenclature, London*, **9** (16), 251–266.

ICHIKAWA, K. 1958. Zur Taxionomia und Phylogenie der triadschen 'Pteriidae' (Lamellibranchia). *Palaeontographica, Abteilung A*, **3**, 131–212.

IREDALE, T. 1939. Great Barrier Reef Expedition, 1928–29: Mollusca, Part I. *Scientific Reports, British Museum (Natural History), London*, **5**, 209–425.

IRWIN, H., CURTIS, C. D. and COLEMAN, M. L. 1977. Isotopic evidence for source of diagenetic carbonates formed during burial of organic-rich sediments. *Nature*, **269**, 209–213.

IVIMEY-COOK, H. C. 1974. The Permian and Triassic deposits of Wales. 295–321. *In* OWEN, T. R. (ed.). *The Upper Palaeozoic and post-Palaeozoic rocks of Wales*. University of Wales Press, Cardiff, viii+426 pp.

—— 1975. The stratigraphy of the Rhaetic and Lower Jurassic in East Antrim. *Bulletin of the Geological Survey of Great Britain*, **50**, 51–69.

—— and POWELL, J. H. 1991. Late Triassic and Early Jurassic biostratigraphy of the Felixkirk Borehole, North Yorkshire. *Proceedings of the Yorkshire Geological Society*, **48**, 367–374.

—— WARRINGTON, G., WORLEY, N. E., HOLLOWAY, S. and YOUNG, B. 1995. Rocks of Late Triassic and Early Jurassic age in the Carlisle Basin, Cumbria (north-west England). *Proceedings of the Yorkshire Geological Society*, **50**, 305–316.

JAEKEL, O. M. J. 1889. Die Selachier aus dem Muschelkalk Lothringens. *Abhandlungen zur geologischen Specialkarte von Elsass-Lothringen*, **3**, 273–332.

—— 1907. *Placochelys placodonta* aus der Obertrias des Bakony Waldes. *Resultate der Wissenschaftlichen Erforschung des Balatonsees. 1, Anhang*: Palaeontologie der Umgebung des Balatonsees, Band 3, Art. 8, 90 pp.

JAIN, S. L. 1983. A review of the genus *Lepidotes* (Actinopterygii: Semionotiformes) with special reference to the species from Kota Formation (Lower Jurassic), India. *Journal of the Palaeontological Society of India*, **28**, 7–42.

JAMES, U. P. 1879. Description of new species of fossils and remarks on some others, from the Lower and Upper Silurian rocks of Ohio. *The Paleontologist*, **3**, 17–24.

JANVIER, P. 1996. *Early vertebrates*. Clarendon Press, Oxford, 393 pp.

JERVEY, M. T. 1988. Quantitative geological modelling of siliciclastic rock sequences and their seismic expression. *In* WILGUS, C. K., HASTINGS, B. S., KENDALL, C. G. ST C., POSAMENTIER, H. G., ROSS, C. A. and VAN WAGONER, J. C. (eds). Sea-level changes: an integrated approach. *Special Publication of the Society of Economic Palaeontologists and Mineralogists*, **42**, 47–69.

JOHNSON, A. L. A. 1984. The palaeobiology of the bivalve families Pectinidae and Propeamussiidae in the Jurassic of Europe. *Zitteliana*, **11**, 3–235.

JOLY, H. 1907. Les fossiles du Jurassique de la Belgique avec description stratigraphique du chaque étage. Première Partie, Infra Lias. *Mémoires du Musée Royal d'Histoire Naturelle de Belgique*, **5**, 1–156.

JONES, T. R. 1863. A monograph on the fossil Estheriae. *Monograph of the Palaeontographical Society*, **14** (62), i–x, 1–134, pls 1–50.

—— 1894. On the Rhaetic and some Liassic Ostracoda of Britain. *Quarterly Journal of the Geological Society, London*, **50**, 156–169.

JUDD, J. W. 1878. The Secondary rocks of Scotland. III. The strata of the western coast and islands. *Quarterly Journal of the Geological Society, London*, **34**, 660–743.

JUKES-BROWNE, A. J. 1885. The geology of the south-west part of Lincolnshire. *Memoirs of the Geological Survey England and Wales*, 1–180.

—— 1912. *The student handbook of stratigraphical geology*. 2nd edition. Stanford, London, 668 pp.

KEEN, A. M. 1951. Outline of a proposed classification of the pelecypod family Cardiidae. *Minutes of the Conchological Club of Southern California*, 6–8.

—— 1958. *Marine shells of tropical west America*. Stanford University Press, Stanford, California, 624 pp.

—— 1969. Superfamily Veneracea. N670–N690. *In* MOORE, R. C. (ed.). *Treatise on invertebrate paleontology. Part N. Mollusca 6, Bivalvia (2)*. Geological Society of America and University of Kansas Press, Boulder, Colorado, and Lawrence, Kansas, ii+N491–N952.

KELLAWAY, G. A. and WELCH, F. B. A. 1993. Geology of the Bristol district. *Memoir of the British Geological Survey*, i–xii+1–199.

KELLING, G. and MOSHRIFF, M. A. 1977. The orientation of fossil bivalves in a pene-littoral sequence (the Rhaetian of S. Wales). *Journal of Sedimentary Petrology*, **47**, 1342–1347.

KENT, P. E. 1968. The Rhaetic beds. 174–187. *In* SYLVESTER-BRADLEY, P. C. and FORD, T. D. (eds). *The geology of the East Midlands*. Leicester University Press, Leicester, 400 pp.

—— 1970. Problems of the Rhaetic in the East Midlands. *Mercian Geologist*, **3**, 361–373.

KERMACK, D. 1984. New prosauropod material from South Wales. *Zoological Journal of the Linnean Society*, **82**, 101–117.

KIDWELL, S. 1989. Stratigraphic condensation of marine transgressive records: origin of major shell deposits in the Miocene of Maryland. *Journal of Geology*, **97**, 1–24.

—— 1991. Condensed deposits in siliciclastic sequences: expected and observed features. 682–695. *In* EINSELE, G., RICKEN, W. and SEILACHER, A. (eds). *Cycles and events in stratigraphy*. Springer-Verlag, Berlin, xix+955 pp.

—— FÜRSICH, F. T. and AIGNER, T. 1986. Conceptual framework for the analysis and classification of fossil concentrations. *Palaios*, 1, 228–238.

KING, W. 1850. A monograph of the Permian fossils of England. *Monograph of the Palaeontographical Society*, **3** (5), i–xxxviii, 1–258, pls 1–28.

KIRBY, W. 1815. *An introduction to entomology: or elements of the natural history of insects...* Vol. 1. London.

KLEIN, E. E. 1885. Beiträge zur Bildung des Schädels der Knochenfische. *Jahreshefte des Vereines für Vaterländische Naturkunde Württemberg*, **41**, 107–261.

KLIE, W. 1938. Krebstiere oder Crustacea III: Ostracoda Muschelkrebse. *Die Tierwelt Deutschlands*, **34**, 1–230.

KOBAYASHI, T. 1954. Studies on the Jurassic trigonians in Japan. Part 1. *Japanese Journal of Geology and Geography*, **25**, 61–80.

KOKEN, E. 1889. Über die Entwicklung der Gastropoden vom Cambrium bis zur Trias. *Neues Jahrbuch für Mineralogie, Geognosie, Geologie und Petrefaktenkunde*, **6**, 305–584.

—— 1892. Über die Gastropoden der rothen Schlernschichten. *Neues Jahrbuch für Mineralogie, Geologie und Paläontologie*, **2**, 25–36.

KONINCK, I. G. de 1881. Faune du calcaire carbonifère de la Belgique, 3ᵉ partie, Gastéropodes. *Annales du Musée royal d'Histoire naturelle de Belgique*, **6**, 1–170.

KOROBKHOV, I. A. 1954. *Spravochnik i metodicheskoe Rukovodstvo po tretichnym mollyuskam Plastinchatozhabernye.* [Handbook on and systematic guide to the Tertiary Mollusca, Lamellibranchia.] Gosudarstvennyi Nauchno-tekhnicheskoi Issledovatel'skii Neftianoi Institut i Gorno-toplivnoi Lit-ry, Leningradskoi Otdelenie, Moskow. 444 pp. [In Russian].

KOZUR, H. 1989. Significance of events in conodont evolution for the Permian and Triassic stratigraphy. *Courier Forschunginstitut Senckenberg*, **117**, 385–408.

—— and MOCK, R. 1974. Zwei neue Conodonten-Arten aus der Trias des Slowakischen Karstes. *Časopis Mineraloggii a Geologii, Prague*, **19**, 135–139.

—— —— 1991. New Middle Carnian and Rhaetian conodonts from Hungary and the Alps. Stratigraphic importance and tectonic implications for the Buda Mountains and adjacent areas. *Jahrbuch der Geologischen Bundesanstalt, Wien*, **134**, 271–297.

KRESIA, R. D. 1981. Storm generated sedimentary structures in subtidal marine facies with samples from the Middle and Upper Ordovician of south western Virginia. *Journal of Sedimentary Petrology*, **51**, 823–848.

—— and BAMBACH, R. K. 1982. The role of storm processes in Palaeozoic shelf seas. 200–208. *In* EINSELE, G. and SEILACHER, A. (eds). *Cyclic and event stratification.* Springer-Verlag, Berlin, xiv+536 pp.

KRISTAN-TOLLMANN, E. 1971. Weitere Beobachtungen an skulptierten Bairdiidae (Ostracoda) der alpinen Trias. *Neues Jarhbuch für Geologie und Paläontologie, Abhandlungen*, **139**, 57–87.

—— and HASIBUAN, F. 1990. Ostracoden aus der Obertrias von Misol (Indonesien). *Mitteilungen der Osterreichischen Geologischen Gesellschaft*, **82**, 173–181.

—— TOLLMANN, A. and HAMEDANI, A. 1980. Beitrage zur kenntnis der Trias von Persien. *Mitteilungen der Osterreichischen Geologischen Gesellschaft*, **73**, 163–235.

KUHN, O. 1934. Ichthyosauria. *Fossilium Catalogus*. I: Animalia, 63, 75 pp.

—— 1949. *Lehrbuch der Paläeozoologie.* E. Schweizerbart, Stuttgart, v+326 pp.

—— 1961. *Die Familien der rezenten und fossilen Amphibien und Reptilien.* Verlagshaus Meisenbach KG, Bamberg, 79 pp.

KUHN-SCHNYDER, E. 1980. Observations on temporal openings of reptilian skulls and the classification of reptiles. 153–175. *In* JACOBS, L. L. (ed.). *Aspects of vertebrate history.* Museum of Northern Arizona Press, Flagstaff, xix+407 pp.

KÜHNE, W. G. 1946. The geology of the fissure-filling 'Holwell 2'; the age determination of the mammalian teeth therein; and a report on the technique employed when collecting teeth of *Eozostrodon* and Microcleptidae. *Proceedings of the Zoological Society, London*, **116**, 729–733.

KUMPERA, O., PRANTL, F. and RUZICKA, B. 1960. Revision of the Nuculanidae from the Ostrava-Karviná District. *Národniho Musea Praze, Sborník (Acta Musei Nationalis Prague)*, **16B**, 17–122.

LAMARCK, J. P. B. A. De M. De 1799. Prodrome d'une nouvelle classification des coquilles, comprenant une rédaction appropriée des caractères génériques, et l'établissement d'un grand nombre de genres nouveaux. *Mémoires de la Société d'Histoire Naturelle, Paris*, **1**, 63–91.

—— 1801. *Système des animaux sans vertèbres, ou table général des classes, des ordres et des genres de ces animaux*. Deterville, Paris, viii+432 pp.

—— 1809. *Philosophie zoologique, ou exposition des considérations relatives à l'histoire naturelle des animaux, la diversité de leur organisation et des facultés qu'ils en obtiennent, aux causes physiques qui main tiennent en eux la vie, et donnent lieu aux mouvements qu'ils exécutent; enfin, à celles qui produisent les unes les sentiments, et les autres l'intelligence de ceux qui en sont doués*. Dentu, Paris. Vol. 1, 422 pp.; vol. 2, 473 pp.

—— 1812. *Extrait du cours de zoologie du Muséum d'Histoire Naturelle sur les animaux invertébres*. Paris, 127 pp.

—— 1818–19. *Histoire naturelle des animaux sans vertèbres, présentánt les caractères généraux et particuliers des ces animaux, leur distribution, leurs classes, leurs familles, leurs genres, et la citation des principales espèces qui s'y rapportent, etc*. Vol. 5, 612 pp; Vol. 6, 343 pp. Published by the author, Paris.

LAMOUROUX, J. V. F. 1821. *Exposition mèthodique des genres de l'ordre des Polypiers, avec leur description et celles des principales éspèces, figurées dans 84 planches: les 63 prémières appartenant a l'Histoire naturelle des Zoophytes d'Ellis et Solander*. Paris, viii+115pp.

LAMPLUGH, G. W., GIBSON, W., WEDD, C. B., SHERLOCK, R. L. and SMITH, B. 1909. The geology of the Melton Mowbray district and south-east Nottinghamshire. *Memoirs of the Geological Survey England and Wales*, 1–118.

—— KILROE, J. R., MCHENRY, A., SEYMOUR, H. J., WRIGHT, W. B. and MUFF, H. B. 1904. The geology of the country around Belfast. *Memoirs of the Geological Survey of Ireland*, 1–166.

LATREILLE, P. A. 1802–03. *Histoire naturelle, générale et párticuliére, des crustacés et des insectes*. Vol. 3. F. Dufart, Paris, 468 pp.

—— 1806–07. *Genera Crustaceorum et Insectorum*. Tomus 1 (1806), Tomus 2 (1807). Paris, 303 pp.

—— 1810. *Considérations générales sur l'ordre naturel des Animaux composant les classes des Crustacés, des Arachnides, et des Insectes: avec un Tableau méthodique de leurs genres, disposés en familles*. Paris, 444 pp.

—— 1817. Les crustacés, les arachnides et les insectes. *In* CUVIER, G. *Le règne animal distribué d'après son organisation, pour servir de base à l'histoire naturelle des animaux et d'introduction à l'anatomie comparée*, edit. 1, Vol. 3. Déterville, Paris, xxix+653 pp.

LEACH, W. E. 1815. *Malacostraca Podophthalmata Britannaeor: descriptions of such British species of the Linnean genus cancer as have their eyes elevated on footstalks*. Nos 1–6. James Sowerby, London.

LEE, G. W. 1920. The Mesozoic rocks of Applecross, Raasay and north-east Skye, with contributions and an appendix by S. S. Buckman. *Memoirs of the Geological Survey of Great Britain*, 1–93.

—— and BAILEY, E. B. 1925. The pre-Tertiary geology of Mull, Loch Aline, and Oban. *Memoirs of the Geological Survey of Great Britain*, 1–140.

—— and PRINGLE, J. 1932. A synopsis of the Mesozoic rocks of Scotland. *Transactions of the Geological Society of Glasgow*, **19**, 158–224.

LEE, J. E. 1872. *In: Report of the British Association for the Advancement of Science*, **1871**, 116.

LEES, G. M. and COX, P. T. 1937. The geological basis of the present search for oil in Great Britain by the D'Arcy Exploration Company, Ltd. *Quarterly Journal of the Geological Society, London*, **93**, 156–194.

—— and TAITT, A. H. 1946. The geological results of the search for oilfields in Great Britain. *Quarterly Journal of the Geological Society, London*, **101**, 255–317.

LESKE, N. G. 1778. *Jacobi Theodori Klein naturalis dispositio Echinodermatum. Edita et descriptionbus novisque inventis et synonymis auctorum aucta.* Leipzig, 278 pp.

LESUER, P. and TASTET, J. P. 1994. Facies, internal structures and sequence of modern Gironde derived muds on the Aquitaine inner shelf, France. *Marine Geology*, **120**, 267–290.

LEYMERIE, A. 1838. Mémoire sur la partie inférieure du système secondaire du Département du Rhône. *Mémoires de la Société Géologique de France, 1ème Série*, **3**, 313–378.

LINDER, F. 1945. Affinities within the Branchiopoda with notes on some dubious fossils. *Arkiv Zoologi*, **37A**, 1–28.

LINDSTRÖM, M. 1970. A suprageneric classification of the conodonts. *Lethaia*, **3**, 427–445.

LINNAEUS, K. 1758. *Systema naturae per tria regna naturae, secundum classes, ordines, genera, species, cum characteribus, differentiis, synonymis, locis.* 10th edition. Laurentius Salvius, Stockholm, ii +824 pp.

LJUBIMOVA, P. S. 1955. Ostrakody Neozoyskikh otlozheniy Srednego Povolzhya i Obshchego Syrta. *Trudy Vsesoyuznogo Neftyanogo Nauchno-Issledovatelskogo Geologorazvedochnogo Instituta (Vnigri), Novaya Seriya*, **84**, 3–189. [In Russian].

LLOYD-MORGAN, C. and REYNOLDS, S. H. 1908. Sketch of the geological history of the Bristol district. *Proceedings of the Bristol Naturalists' Society, Series 4*, **2**, 5–26.

LOEBLICH, A. R. and TAPPAN, H. 1961. Suprageneric classification of the Rhizopodea. *Journal of Paleontology*, **35**, 245–330.

—— —— 1984. Suprageneric classification of the Foraminiferida (Protozoa). *Micropalaeontology*, **30**, 1–70.

LORD, A. R. and BOOMER, I. D. 1990. The occurrence of ostracods in the Triassic/ Jurassic boundary interval. *Les Cahiers de l'Université Catholique de Lyon, Série Sciences*, **3**, 119–126.

—— and BOWN, P. R. (eds). 1987. *Mesozoic and Cainozoic stratigraphical micropalaeontology of the Dorset coast and Isle of Wight, southern England.* British Micropalaeontological Society, Field Guide No. 1, 183 pp.

LOUTIT, T. S., HARDENBOL, J., VAIL, P. R. and BAUM, G. R. 1988. Condensed sections: the key to age dating and correlation of continental margin sequences. *In* WILGUS, C. K., HASTINGS, B. S., KENDALL, C. G. St C., POSAMENTIER, H. G., ROSS, C. A. and VAN WAGONER, J. C. (eds). Sea-level changes: an integrated approach. *SEPM Special Publication*, **42**, 183–216.

LUNDGREN, B. 1878. *Studier öfver Faunan i den stenkolsförande Formationen i nordvästra Skåne. Minnesskrift utgifven af Kongliga Fysiografiska Sällskapet i Lund med Anledning af dess Hundraårsfest den 3 October 1878.* Lund, 57 pp.

LYDEKKER, R. 1888. *Catalogue of the fossil Reptilia and Amphibia in the British Museum (Natural History): part 1. Containing the orders Ornithosauria, Crocodilia, Dinosauria, Squamata, Rynchocephalia, and Proterosauria.* British Museum (Natural History), London, xxviii+309 pp.

—— 1889. *Catalogue of the fossil Reptilia and Amphibia in the British Museum: part II. Ichthyopterygia and Sauropterygia.* British Museum (Natural History), London, xxiii+307 pp.

—— 1890. *Catalogue of the fossil Reptilia and Amphibia in the British Museum: part IV. Anomodontia, Ecaudata, Caudata, and Labyrinthodontia; and supplement.* British Museum (Natural History), London, xxiii+295 pp.

LYMAN, T. 1865. *Illustrated catalogue of the Museum of Comparative Zoology at Harvard College. No. 1. Ophiuridae and Astrophytidae.* Cambridge, Mass., 200 pp.

MACFADYEN, W. A. 1970. *Geological highlights of the West Country – a Nature Conservancy Handbook.* Butterworths, London, 296 pp.

MACQUAKER, J. H. S. 1994. Palaeoenvironmental significance of 'bone-beds' in organic-rich mudstone successions: an example from the Upper Triassic of south-west Britain. *Zoological Journal of the Linnean Society*, **112**, 285–308.

—— and TAYLOR K. G. 1996. A sequence stratigraphic interpretation of a mudstone-dominated succession: the Lower Jurassic Cleveland Ironstone Formation, U.K. *Journal of the Geological Society, London*, **153**, 759–770.

—— FARRIMOND, P. and BRASSELL, S. C. 1986. Biological markers in the Rhaetian black shales of south west Britain. *Organic Geochemistry*, **10**, 93–100.

—— —— YOUNG, T. P. and CURTIS, C. D. 1996. Sedimentological and geochemical controls on ooidal ironstone and 'bone-bed' formation and some comments on their sequence stratigraphic significance. *In* HESSELBO, S. and PARKINSON, D. N. (eds). Sequence stratigraphy in British geology. *Special Publication of the Geological Society, London*, **103**, 97–107.

MAGOR, P. M. 1978. Rhaetic trace fossils attributable to the ichnogenus *Kouphichnium* Nospca. *Circular of the Palaeontological Association*, **94a**, 7 (abstract).

MAISEY, J. G. 1975. The interrelationships of phalacanthous selachians. *Neues Jahrbuch für Geologie und Paläontologie, Monatshefte*, **9**, 553–567.

—— 1977. The fossil selachian fishes *Palaeospinax* Egerton, 1872 and *Nemacanthus* Agassiz, 1837. *Zoological Journal of the Linnean Society*, **60**, 259–273.

—— 1982. The anatomy and interrelationships of Mesozoic hybodont sharks. *American Museum Novitates*, **2724**, 1–48.

MANNING, P. I. and WILSON, H. E. 1975. The stratigraphy of the Larne borehole, county Antrim. *Bulletin of the Geological Survey of Great Britain*, **50**, 1–50.

—— ROBBIE, J. A. and WILSON, H. E. 1970. Geology of Belfast and the Lagan Valley. *Memoirs of the Geological Survey of Northern Ireland*, 1–242.

MANTELL, G. A. 1844. *The Medals of Creation; or, first lessons in geology and the study of organic remains.* 2 vols. London.

MARSH, O. C. 1880. Notice of Jurassic mammals representing two new orders. *American Journal of Science*, **20**, 235–239.

—— 1881. Classification of the Dinosauria. *American Journal of Science*, **23**, 81–86.

—— 1885. On the classification and affinities of the Dinosauria of Europe and America. *Report of the British Association for the Advancement of Science*, **1884**, 763–765.

MARSHALL, J. E. A. and WHITESIDE, D. I. 1980. Marine influences in the Triassic 'uplands'. *Nature*, **287**, 627–628.

MARTILL, D. M. and DAWN, A. 1986. Fossil vertebrates from new exposures of the Westbury Formation (Upper Triassic) at Newark, Nottinghamshire. *Mercian Geologist*, **10**, 127–133.

MARTIN, J. 1860. Paléontologie stratigraphique de l'Infra Lias du département de la Côte-d'Or. *Mémoires de la Société Géologique de France, 2ème Serie*, **7**, 1–100.

MARWICK, J. 1953. Division and faunas of the Hokonui System (Triassic and Jurassic). *Palaeontological Bulletin of the New Zealand Geological Survey*, **21**, 1–141.

MATSUMOTO, H. 1915. A new classification of the Ophiuroidea, with descriptions of new genera and species. *Proceedings of the Academy of Natural Sciences, Philadelphia*, **67**, 43–92.

MAYALL, M. J. 1981. The late Triassic Blue Anchor Formation and the initial Rhaetian marine transgression in south-west Britain. *Geological Magazine*, **118**, 377–384.

—— and WRIGHT, V. P. 1981. Algal tuft structures in the stromatolites from the Upper Triassic of south-west England. *Palaeontology*, **24**, 655–660.

MCALLISTER, J. A. 1985. Re-evaluation of the formation of spiral coprolites. *University of Kansas Paleontological Contributions*, **114**, 1–12.

MCCOY, F. 1844. *A Synopsis of the characters of the Carboniferous Limestone Fossils of Ireland [with a 'Notice' by Sir R. Griffith.].* Dublin, viii+207 pp. [Privately published by Sir R. J. Griffith. Subsequently Williams and Norgate re-issued the work with a new title-page dated 1862].

MCGOWAN, C. 1989. *Leptopterygius tenuirostris* and other long-snouted ichthyosaurs from the English Lower Lias. *Palaeontology*, **32**, 409–427.

—— 1996. The taxonomic status of *Leptopterygius* von Huene, 1922 (Reptilia: Ichthyosauria). *Canadian Journal of Earth Sciences*, **33**, 439–443.

MEEK, F. B. 1864. Check list of the invertebrate fossils of North America. Cretaceous and Jurassic. *Smithsonian Miscellaneous Collections*, **7**, 1–40.

—— 1865. Description of fossils from the auriferous slates of California. *Geological Survey of California, Geology*, **1**, 477–482.

—— and HAYDEN, F. V. 1861. Description of new Lower Silurian (Primordial), Jurassic, Cretaceous and Tertiary fossils collected in Nebraska, etc. *Proceedings of the Academy of Natural Sciences of Philadelphia*, **10**, 41–59.

MEHÉS, G. 1911. Über Trias-Ostrakoden aus dem Bakony. *Resultate der Wissenschaftlichen Erforschung des Balatonsees, Anhang zu Band 1, Teil 1: Paläontologie der Umgebung des Balatonsees*, **3**, 1–38, 4 taf. Wien.

282   *Fossils of the Rhaetian Penarth Group*

MELVILLE, R. V. 1956. The stratigraphical palaeontology, ammonites excluded, of the Stowell Park Borehole. *Bulletin of the Geological Survey of Great Britain*, **11**, 67–139.

MENKE, C. T. 1828. *Synopsis methodica Molluscorum generum omnium et specierum earum quae in Museo Menkeano adservatur.* Pyrmonti, 91 pp.

MERIAN, P. 1852. Ueber das Vorkommen der St. Cassianformation in den Bergamasker Alpen und in der Kette des Rhätikon. *Bericht ueber die Verhandlungen der Naturforschenden Gesellschaft in Basel*, **10**, 147–150.

MEYER, H. von 1849. Fossile Fische aus dem Muschelkalk von Jena, Querfurt und Esperstädt. *Palaeontographica*, **1**, 195–208.

—— 1867. *Psephoderma anglicum* aus dem Bone Bed in England. *Palaeontographica*, **15**, 261–263.

MICHELSON, O. 1975. Lower Jurassic biostratigraphy and ostracods of the Danish Embayment. *Danmarks Geologiske Undersogelse, 2 Raekke*, **104**, 1–289.

MILNE-EDWARDS, H. 1848. Note sur la classification naturelle des mollusques gastéropodes. *Annales des Sciences naturelles Zoologie, Series 3*, **9**, 102–112.

—— and HAIME, J. 1848. Recherches sur les polypes. *Annales de Science Naturelles, Paris*, **9**, 37–89, 211–344; **10**, 65–114, 209–320.

MONTFORT, D. de 1808. *Conchyliologie systématique et classification méthodique des coquilles.* Vol. 1. F. Schoell, Paris, lxxxvii+409 pp.

MOORE, C. 1859. On Triassic beds near Frome and their organic remains. *Report of the British Association for the Advancement of Science*, **1858**, 93–94.

—— 1861a. On the contents of three square yards of Triassic drift. *Report of the British Association for the Advancement of Science*, **1860**, 87–88.

—— 1861b. On the zones of the Lower Lias and the *Avicula contorta* Zone. *Quarterly Journal of the Geological Society, London*, **17**, 483–516.

—— 1867. On abnormal conditions of secondary deposits when connected with the Somersetshire and South Wales coal-basin; and on the age of the Sutton and Southerndown Series. *Quarterly Journal of the Geological Society, London*, **23**, 449–568.

—— 1881. On abnormal geological deposits in the Bristol district. *Quarterly Journal of the Geological Society, London*, **37**, 67–82.

MOORE, R. C. (ed.). 1960. *Treatise on invertebrate palaeontology. Part 1. Mollusca 1.* Geological Society of America and University of Kansas, Boulder, Colorado, and Lawrence, Kansas, xxiii+351 pp.

MORRIS, J. 1843. *Catalogue of British fossils.* J. Van Voorst, London, 222 pp.

—— and LYCETT, J. 1853. A monograph of the Mollusca from the Great Oolite, chiefly from Minchinhampton and the Coast of Yorkshire. Part 2, Bivalves. *Monograph of the Palaeontographical Society*, **7** (23), 1–80, pls 1–8.

MORTENSEN, T. 1939. *A monograph of the Echinoidea.* Vol. 3, no. 1. *Aulodonta.* Oxford University Press, London, 370 pp.

MÜLLER, G. W. 1894. *Die Ostracoden des Golfes von Neapel und der angrenzenden Meeres-abschnitte, Herausgegeben von der Zoologischen Station zu Neapel*: 21. Berlin, i–vii, 1–404.

MÜLLER, J. 1846a. Über den Bau und die Grenzen der Ganoiden und über das natürliche System der Fische. *Physikalisch-Mathematische Abhandlungen der koeniglichen Akademie der Wissenschaften zu Berlin, Jahrgang 1844*, 117–216.

—— 1846*b*. Fernere Bemerkungen über den Bau der Ganoiden. *Bericht der Deutschen Akademie der Wissenschaften zu Berlin, Jahrgang 1846*, 67–85.

—— and TROSCHEL, F. H. 1840. Über die Gattungen der Ophiuren. *Archiv für Naturgeschichte, Sechster Jahrgang, Erster Band, Zoologie*, 326–330.

MÜNSTER, G. 1833. *Verzeichniss der Versteinerungen, welche in der Kreis-Naturlischen Sammlung zu Bayreuth vorhanden sind*. Beyreuth, 115 pp.

—— 1839. Decapoda macroura. Abbildung und Beschreibung der fossilen langschwänzigen Krebse in Kalkschiefern von Bayern. *In* DUNKER, W. B. R. H. 1839–46. *Beiträge zur Petrefacten-Kunde*, Hft 2. Beyreuth.

NEUMAYER, L. 1904. Die Koprolithen des Perms von Texas. *Palaeontographica*, **51**, 121–128.

NEVILLE, R. S. W. 1967. The Rhaetic (Part 1 of the Bonebed). *Journal of the University of Sheffield Geological Society*, **5**, 33–37.

NEWELL, N. D. 1957. Notes on certain primitive heterodont pelecypods. *American Museum of Natural History Novitates*, **1857**, 1–14.

—— 1965. Classification of the Bivalvia. *American Museum of Natural History Novitates*, **2206**, 1–25.

—— 1969. Classification of the Bivalvia. N205–N218. *In* MOORE, R. C. (ed.). *Treatise on invertebrate paleontology. Part N. Mollusca 6, Bivalvia (1)*. Geological Society of America and University of Kansas Press, Boulder, Colorado, and Lawrence, Kansas, xxxviii+N1–N489.

NEWMAN, W. A., ZULLO, V. A. and WITHERS, T. H. 1969. Cirripedia. R206–R295. *In* MOORE, R. C. (ed.). *Treatise on invertebrate paleontology. Part R, Arthropoda 4*. Geological Society of America and University of Kansas Press, Boulder, Colorado, and Lawrence, Kansas, xxxvi+398 pp.

NEWTON E. T. 1899. On a megalosauroid jaw from Rhaetic beds near Bridgend (Glamorganshire). *Quarterly Journal of the Geological Society, London*, **55**, 89–96.

NEWTON, R. B. 1891. *Systematic list of the Frederick E. Edwards Collection of the British Oligocene and Eocene Mollusca in the British Museum*. British Museum, London, xxviii+365 pp.

NICHOLSON, H. A. 1873. Contributions to the study of the errant annelides of the older Palaeozoic rocks. *Proceedings of the Royal Society, London*, **21**, 288–290.

NILSSON, S. 1832. Djur-petrifikater funna i Skånes Stenkolsbildning. *Kungliga Svenska Vetenskaps Akademien, Stockholm, Handlingar*, 352–354.

NOPCSA, F. 1923. Die Familien der Reptilien. *Fortscritte der Geologie und Paläontologie, Berlin*, **2**, i–iii+1–210.

—— 1928. The genera of reptiles. *Palaeobiologica*, **1**, 163–188.

OLD, R. A., SUMBLER, M. G. and AMBROSE, K. 1987. Geology of the country around Warwick. *Memoir of the British Geological Survey*, i–viii+1–93.

OLIVIER, G.-A. 1789–1808. *Entomologie, ou Histoire naturelle des Insectes: Coléoptères*. 6 vols+Atlas. Paris.

ORBIGNY, A. C. V. M. d'. 1826. Tableau méthodique de la classe des Céphalopodes. *Annales des Sciences Naturelles, Paris, 1ème Série*, **7**, 245–314.

—— 1839. Foraminifères. *In* SAGRA, R. de la. *Histoire physique, politique et naturelle de l'île de Cuba*. Paris, xlviii+224 pp.

—— 1842. *Voyage dans l'Amérique méridionale (le Brésil, la République oriental de l'Uruquay, la République Argentine, la Patagonie, la République du Chili, la*

République de Bolivia, la République du Péron) exécuté pendant les annees 1826, 1827, 1829, 1830, 1831, 1832, et 1833. Vol. 3, pt 4 (Paléontologie). Pitois-Levrault, Paris, 188 pp.

—— 1844. *Paléontologie française terrains crétacés.* Vol. 3. *Mollusques.* G. Masson, Paris, 807 pp.

—— 1849. *Prodrome de paléontologie stratigraphique universelle des animaux mollusques et rayonnés faisant suite au cours élémentaire de paléontologie et de géologie stratigraphiques.* Vol. 1. V. Masson, Paris, ix+394 pp.

—— 1852. *Cours élémentaire de Paléontologie et de Géologie stratigraphiques.* Vol. 2, part 1. Paris.

OSBORN, H. F. 1903. On the primary division of the Reptilia, into two sub-classes, Synapsida and Diapsida. *Science, New Series*, **17**, 275–276.

OWEN, E. 1754. *Observations on the Earths, Rocks, Stones, and Minerals, for some miles about Bristol, and on the nature of the Hot-well, and the virtues of its waters.* London, 250 pp.

OWEN, R. 1840. Report on British fossil reptiles. *Report of the British Association for the Advancement of Science*, **1839**, 43–126.

—— 1842. Report on British fossil reptiles. *Report of the British Association for the Advancement of Science*, **1841**, 60–204.

—— 1846. On the supposed fossil bones of birds from the Wealden. *Quarterly Journal of the Geological Society, London*, **2**, 96–102.

—— 1859. Palaeontology. 91–176. *In Encyclopedia Britannica.* 8th edition, vol. 17. Edinburgh.

—— 1860. On the order of fossil and recent Reptilia, and their distribution in time. *Report of the British Association for the Advancement of Science*, **1859**, 153–166.

—— 1871. Fossil Mammalia of the Mesozoic formations. *Monograph of the Palaeontographical Society*, **24** (110), i-vi, 1–115, pls 1–40.

PAALZOW, R. 1935. Die Foraminiferan im Zechstein des östlichen Thüringen. *Jahrbuch der Preussischen geologischen Landesanstalt*, **56**, 26–45.

PACKARD, A. S. 1886. A new arrangement of the orders of insects. *American Naturalist*, **20**, 808.

PALMER, C. P. 1975. The British Lower Jurassic species of the bivalve genus Cardinia. *Bulletin of the British Museum (Natural History), Geology Series*, **26**, 3–44.

PANDER, C. H. 1856. *Monographie der fossilen Fische des Silurischen Systems der Russisch-Baltischen Gouvernements.* Kaiserlichen Akademie der Wissenschaften, St Petersberg, 91 pp.

PARKER, W. K. and JONES, T. R. 1859. On the nomenclature of the Foraminifera. Part 2. On the species enumerated by Walker and Montagu. *Annals and Magazine of Natural History, Series 3*, **4**, 333–351.

PATTERSON, C. 1965. The phylogeny of the chimaeroids. *Philosophical Transactions of the Royal Society of London, Series B*, **249**, 101–219.

PATTISON, J. 1981. Permian. 70–77. *In* JENKINS, D. G. and MURRAY, J. W. (eds). *Stratigraphical atlas of fossil foraminifera.* 1st edition. Ellis Horwood Ltd for The British Micropalaeontological Society, Chichester, 310 pp.

PEACH, B. N., GUNN, W. and NEWTON, E. T. 1901. On a remarkable volcanic vent of Tertiary age in the island of Arran, enclosing fossiliferous Mesozoic rocks. *Quarterly Journal of the Geological Society, London*, **57**, 226–243.

PENNANT, T. 1776–77. *British zoology: illustrated by plates and brief explanations.* 4 vols. Warrington, London.

PERSOON, C. H. 1805–07. *Synopsis Plantarum, seu Enchiridium Botanicum, complectens enumerationem systematicam specierum hucusque cognitarum.* Paris.

PFLÜCKER y RICO, L. 1868. Das Räth (die Rhätische Gruppe) in der umgegend von Göttingen. *Zeitschrift der Deutschen geologischen Gesellschaft,* **20**, 397–432.

PHILLIPS, J. 1871. *Geology of Oxford and the valley of the Thames.* Clarendon Press, Oxford, xxiv+529 pp.

PILSBRY, H. A. 1916. The sessile barnacles (Cirripedia) contained in the collections of the United States National Museum, including a monograph of the American species. *Bulletin of the United States National Museum,* **93**, 1–366.

PLIENINGER, T. 1847. Abbildungen von Zahnen aus der obern Grenzebreccie der Keupers bei Degerloch und Steinnebronnen. *Jahreshefte des Vereins für vaterländische Naturkunde in Württemberg,* **3**, 164–167.

PLUMMER, H. J. 1945. Smaller foraminifera in the Marble Falls, Smithwick and lower Strawn strata around the Llano uplift in Texas. *Publication of the University of Texas,* **4401**, 209–271.

POEL, L. van de 1959. Faune malacologique du Hervien. *Bulletin de l'Institut Royal des Sciences Naturelles de Belgique,* **35**, No. 15 (1–26), No. 16 (1–28).

POJETA, J. Jr 1978. The origin and early taxonomic diversification of pelecypods. *Philosophical Transactions of the Royal Society of London, Series B,* **284**, 225–246.

POMEL, A. 1883. *Classification méthodique et genera des échinides vivants et fossiles.* Adolphe Jourdan, Alger, 131 pp.

PONDER, W. F. and WARÉN, A. 1988. Classification of the Caenogastropoda and Heterostopha – a list of the family group names and higher taxa. *Malacological Review, Supplement,* **4**, 288–326.

POOLE, E. G. 1979. The Triassic-Jurassic boundary in Great Britain. *Geological Magazine,* **116**, 303–311.

—— and WHITEMAN, A. J. 1966. Geology of the country around Nantwich and Whitchurch. *Memoirs of the Geological Survey of Great Britain,* i-viii+1–154.

POPOV, Y. A., DOLLING, W. R. and WHALLEY, P. E. S. 1994. British Upper Triassic and Lower Jurassic Heteroptera and Coleorrhyncha (Insecta: Hemiptera). *Genus,* **5**, 307–347.

PORTLOCK, J. E. 1843. *Report on the Geology of the County of Londonderry, and of parts of Tyrone and Fermanagh.* Milikan and Longmans, Dublin and London, xxxi+784 pp.

PRINGLE, J. 1922. On a boring for coal at Westbury, Wiltshire. *Summary of Progress of the Geological Survey of Great Britain for* **1921**, 146–155.

PURNELL, M. A. and DONOHUE, P. C. J. 1998. Skeletal architecture, homologies and taphonomy of ozarkodinid conodonts. *Palaeontology,* **41**, 57–102.

QUENSTEDT, F. A. 1856–58. *Der Jura.* Parts 1–3 (1856) pp. 1–576; Part 4 (1857) pp. 577–823; Part 5 (1858) title pages and index. H. Laupp, Tübingen, 842 pp.

RAAF, J. F. M. de, BOERSMA, J. R. AND VAN GELDER, J. 1977. Wave generated structures and sequences from a shallow marine succession, Lower Carboniferous, County Cork, Ireland. *Sedimentology,* **24**, 451–483.

RAFINESQUE, C. S. 1815. *Analyse de la Nature ou Tableau de l'Univers et des corps organisés, etc.* Palermo, 224 pp.

RAWSON, P. F., GREENSMITH, J. T. and SHALABY, S. E. 1982. Coarsening-upwards cycles in the uppermost Staithes and Cleveland Ironstone formations (Lower Jurassic) of the Yorkshire Coast, England. *Proceedings of the Geologists' Association*, **94**, 91–93.

RAYMOND, P. E. 1946. The genera of fossil Conchostraca – an order of bivalved Crustacea. *Bulletin of Harvard University Museum of Comparative Zoology*, **96**, 218–307.

REGAN, C. T. 1906. A classification of the selachian fishes. *Proceedings of the Zoological Society of London*, **1906**, 722–758.

—— 1923. The skeleton of *Lepidosteus*, with remarks on the origin and evolution of the lower neopterygian fishes. *Proceedings of the Zoological Society of London*, **1923**, 445–461.

—— 1925. The fishes of the genus *Gigantura*. *Bulletin of the American Museum of Natural History*, **51**, 53–59.

REIF, W.-E. 1976. Sedimentologie und genese von bonebed. *Zentralblat für Geologie und Paläontologie*, **5**, 252–255.

—— 1978. Tooth enameloid as a taxonomic criterion: 2. Is '*Dalatias barnstonensis*' Sykes, 1971 (Triassic, England) a squalomorphic shark? *Neues Jahrbuch für Geologie und Paläontologie, Monatshefte*, **1978**, 42–58.

REUSS, A. E. 1861. Beiträge zur Kenntniss der tertiären Foraminiferan-Fauna. *Sitzungsberichte der Kaiserlichen Akademie der Wissenschaften. Mathematisch-naturwissenschaftliche Classe, Wien*, **42**, 355–370.

—— 1862. Entwurk einer systematischen Zusammenstellung der Foraminiferen. *Sitzungsberichte der Kaiserlichen Akademie der Wissenschaften. Mathematisch-naturwissenschaftliche Classe, Wien*, **44**, 355–396.

REYNOLDS, S. H. 1946. The Aust section. *Proceedings of the Cotteswold Naturalists' Field Club*, **29**, 29–39.

—— and VAUGHAN, A. 1904. The Rhaetic Beds of the South-Wales direct line. *Quarterly Journal of the Geological Society, London*, **60**, 194–214.

RICHARDSON, L. 1901. Mesozoic geography of the Mendip Archipelago. *Proceedings of the Cotteswold Naturalists' Field Club*, **14**, 59–73.

—— 1904. *A handbook to the geology of Cheltenham and neighbourhood.* Norman Sawyer and Co., Cheltenham, xii +303 pp.

—— 1905. The Rhaetic and contiguous deposits of Glamorganshire. *Quarterly Journal of the Geological Society, London*, **61**, 385–424.

—— 1906. On the Rhaetic and contiguous deposits of Devon and Dorset. *Proceedings of the Geologists' Association*, **19**, 401–409.

—— 1911. The Rhaetic and contiguous deposits of West, Mid, and part of East Somerset. *Quarterly Journal of the Geological Society, London*, **67**, 1–74.

—— 1912. The Rhaetic rocks of Warwickshire. *Geological Magazine*, dec. 5, **9**, 24–33.

—— 1933. The country around Cirencester. *Memoirs of the Geological Survey England and Wales*, 1–119.

—— and TUTCHER, J. W. 1916. On *Pteromya crowcombeia* Moore and some species of *Pleuromya* and *Volsella* from the Rhaetic and Lower Lias. *Proceedings of the Yorkshire Geological Society*, **19**, 51–58.

RICHTER, M. and SMITH, M. 1995. A microstructural study of the ganoine tissue of selected lower vertebrates. *Zoological Journal of the Linnean Society*, **114**, 173–212.

RIEBER, H. 1980. Ein Conodonten-cluster aus der Grenzbitumenzone (Mittlere Trias) des Monte San Giorgio (Kt. Tessin/Schweiz). *Annalen des Naturhistorisches Museums, Wien*, **83**, 265–274.

RIEPPEL, O. 1985. Die Triasfauna der Tessiner Kalkalpen. XXV. Die Gattung *Saurichthys* (Pisces, Actinopterygii) aus der mittleren Trias des Monte San Giorgio, Kanton Tessin. *Schweizerische Paläontologische Abhandlungen*, **108**, 1–103.

RILEY, H. and STUTCHBURY, S. 1836. A description of various fossil remains of three distinct saurian animals discovered in the Autumn of 1834, in the Magnesian Conglomerate on Durdham Down, near Bristol. *Proceedings of the Geological Society, London*, **2**, 397–399.

ROBINSON, P. L. 1957. The Mesozoic fissures of the Bristol Channel area and their vertebrate faunas. *Zoological Journal of the Linnean Society*, **43**, 260–282.

RÖDER, H. A. 1882. *Beiträge zur Kenntnis des Terrain à Chailles und seiner Zweischalen in der Umgegend von Pfirt im Ober-Elsass*. Inaugural Dissertation Mathamatische naturwissenschaftelichen Fakultie Kaiser-Wilhelmsuniversität Strassbürg. Schultz, Strasbourg, 110 pp.

RÖDING, P. F. 1798. *Museum Boltenianum sive Catalogus cimeliorum e tribus regnis naturae quae olim collegerat Joa. Fried. Bolten, M.D. p.d., pars secunda*. Typis Johan Christi Trapii (Hamburg), viii+199 pp.

ROMANO, M. and WHYTE, M. A. 1990. *Selenichnites*, a new name for the ichnogenus *Selenichnus* Romano and Whyte, 1987. *Proceedings of the Yorkshire Geological Society*, **48**, 221.

ROMER, A. S. 1955. Fish origins – fresh or salt water? 262–280. *In* BIGELOW, H. B. (ed.). *Papers in marine biology and oceanography*. Pergamon Press, London, xx +498 pp.

ROTHMALER, W. 1951. Die Abteilungen und Klassen der Pflanzen. *Feddes Repertorium Specierum Novarum Vegetabilis*, **54**, 256–266.

ROWELL, A. J. 1965. Inarticulata. H260–H296. *In* MOORE, R. C. (ed.). *Treatise on invertebrate paleontology. Part H. Brachiopoda (1)*. Geological Society of America and University of Kansas Press, Boulder, Colorado, and Lawrence, Kansas, xxxii+H1–H521.

RUTHERFORD, M. M. and AINSWORTH, N. R. 1989. Micropalaeontological and stratigraphical recognition of the Triassic/Jurassic boundary in the North Celtic Sea and Fastnet basins. 45–69. *In* BATTEN, D. J. and KEEN, M. C. (eds). *Northwest European micropalaeontology and palynology*. Ellis Horwood Ltd, Chichester, x+298 pp.

RYLEY, C. C. and FÅHRÆUS, L. E. 1994. Two new genera *Comperniodontella* n. gen. and *Galeodontella* n. gen., and new multielement species of *Chirodella* Hirschmann, 1959 and *Cypridodella* Mosher, 1966 (Conodonta) from the Mamonia Complex (Upper Triassic), Cyprus. *Neues Jahrbuch für Geologie und Paläontologie, Abhandlungen*, **193**, 21–54.

RZEHAK, A. 1888. Die Foraminiferen des kieseligen Kalkes von Nieder-Hollabrunn und des Melettamergels der Umgebung von Bruderndorf in Niederösterreich. *Annalen der Hofmuseum der Naturhistorische, Wien*, 3, 257–270.

SALTER, J. W. 1857. On annelide-burrows and surface markings from the Cambrian rocks of the Longmynd. *Quarterly Journal of the Geological Society, London*, **13**, 199–206.

SALVINI-PLAWÉN, L. 1980. A reconsideration of systematics in the Mollusca (phylogeny and higher classification). *Malacologia*, **19**, 249–278.

—— and HASZPRUNAR, G. 1987. The Vetigastropoda and the systematics of streptoneurous gastropods (Mollusca). *Journal of Zoology*, **A211**, 747–770.

SANDERS, W. 1876. On certain large bones in the Rhaetic Beds at Aust Cliff, near Bristol. *Report of the British Association for the Advancement of Science*, **1875**, 80–81.

SAPORTA, L. C. J. G. de 1872. *Paléontologie française ou description des fossiles de la France (commencée par Alcide d'Orbigny et) continuée par une réunion de paléontologistes. 2 sér. Végétaux. Plantes jurassiques*. Vol. 1. G. Masson, Paris, 432 pp.

—— 1887. Nouveaux documents relatifs aux organismes problématiques des anciennes mers. *Bulletin de la Société Géologiques de France, 3ème Série*, **15**, 286–302.

SARS, G. O. 1866. Oversigt af Norges marine Ostracoder. *Forhandlinger i Videnskabs-Selskabet i Christiania*, **1865**, 1–130.

—— 1867. *Histoire naturelle des Crustacés d'eau douce Norvège*. C. Johnson, Christiania, 145 pp.

—— 1872. Undersøgelser over Hardangerfjkordens Fauna. *Forhandlinger i Videnskabs-Selskabet i Christiania*, **1871**, 246–255.

—— 1888. Nye bidrag til Kundskaben om Middelhavets Invertebratfauna: 4. Ostracoda Mediterranea (Sydeuropaeiske Ostracoder). *Arkiv for Mathematik og Naturvidenskab*, **12**, 173–324.

SAVAGE, R. J. G. 1993. Vertebrate fissure faunas with special reference to Bristol Channel Mesozoic faunas. *Journal of the Geological Society, London*, **150**, 1025–1034.

—— and LARGE, N. F. 1966. On *Birgeria acuminata* and the absence of labyrinthodonts from the Rhaetic. *Palaeontology*, **9**, 135–141.

—— and WALDMAN, M. 1966. *Oligokyphus* from Holwell Quarry, Somerset. *Proceedings of the Bristol Naturalists' Society*, **31**, 185–192.

SCHÄFLE, L. 1929. Über Lias- und Doggeraüstern. *Geologischen und Paleontologischen Abhandlungen, Jena, Neue Folge*, **17**, 65–150.

SCHIMPER, W. P. 1879. *In* ZITTEL, K. A. von, SCHIMPER, W. P. and SCHENK, A. S. 1879–90. *Handbuch der Palaeontologie. 2 Abtheilung: palaeophytologie*. R. Oldenbourg, Munich, xi+958 pp.

SCHLOTHEIM, E. F. von 1820. *Die Petrefactenkunde auf ihrem jetzigen Stanpunkte durch die Beschreibung seiner Sammlung versteiner und fossiler Überreste des Their- und Pflanzenreichs der Vorwelt erläutert*. Becker, Gotha, lxii+437 pp.

SCHLUMBERGER, C. 1890. Seconde note sur les holothuroidées du Calcaire grossier. *Mémoires de la Société Géologique de France, 3ème Série*, **18**, 191–206.

SCHMITZ, B., ABERG, G., WERDELIN, L., FOREY, P. and BENDIX-ALMGREEN, S. E. 1991. 87Sr/86Sr, Na, F, Sr and La in skeletal fish debris as a measure of the palaeosalinity of fossil fish habitats. *Bulletin of the Geological Survey of America*, **103**, 786–794.

SCHNEIDER, G. F. 1956. *In* SCHNEIDER, G. F. *et al.* Materialy po Paleontologii. Novye

Semeystva i Rody: Ostracoda. *Trudy Vsesoyuznogo Nauchno-Issledovatelskogo Geologicheskogo Instituta (Vsegei), Novaya Seriya*, **12**, 87–144, 306–323. [In Russian].

SCHWAGER, C. 1877. Quadro del proposto sistema de classificazione dei foraminiferi con guscio. *Bollettino del R. Comitato Geologico d'Italia*, **8**, 18–27.

SCOPOLI, G. A. 1777. *Introductio ad historiam naturalem sistentes genera lapidum, plantarum et animum, hactenus detecta caracteribus essentialibus donata, in tribus divisa, subinde ad leges naturae*. Prague, 506 pp.

SEELEY, H. G. 1887. On the classification of the fossil animals commonly called Dinosauria. *Proceedings of the Royal Society of London*, **43**, 165–171.

—— 1891. On *Saurodesmus Robertsoni* (Seeley), a crocodilian reptile from the Rhaetic of Linksfield, in Elgin. *Quarterly Journal of the Geological Society, London*, **47**, 166–170.

—— 1898. On large terrestrial saurians from the Rhaetic beds of Wedmore Hill, described as *Avalonia Sanfordi* and *Picrodon Herveyi*. *Geological Magazine*, **5**, 1–6.

SEIGLIE, G. A. and BERMÚDEZ, P. J. 1965. Observaciones sobre foraminiferos rotaliformes con camaras suplementarios o estructuras semejantes (1). *Boletín del Instituto Oceanográfico, Universidad de Oriente*, **4**, 155–171.

SEILACHER, A. 1953a. Studien zur Palichnologie. I. Über die Methoden der Palichnologie. *Neues Jarhbuch für Geologie und Paläontologie, Abhandlungen*, **96**, 421–452.

—— 1953b. Studien zur Palichnologie. II. Die fossilen Ruhespuren (Cubichnia). *Neues Jarhbuch für Geologie und Paläontologie, Abhandlungen*, **98**, 87–124.

—— 1964. Biogenic sedimentary structures. 296–316. *In* IMBRIE, J. and NEWELL, N. D. (eds). *Approaches to paleoecology*. Wiley, New York, viii+432 pp.

—— 1967. Bathymetry of trace fossils. *Marine Geology*, **5**, 413–428.

—— 1982. Distinctive features of sandy tempestites. 333–349. *In* EINSELE, G. and SEILACHER, A. (eds). *Cyclic and event stratification*. Springer-Verlag, Berlin, xiv +536 pp.

SEWARD, A. C. 1904. *The Jurassic Flora*. Vol. II. *Liassic and Oolitic Floras of England. Catalogue of the Mesozoic plants in the Department of Geology, British Museum (Natural History)*. 4. Trustees of the British Museum (Natural History), London, xvi +192 pp.

SHORT, A. R. 1904. A description of some Rhaetic sections in the Bristol district, with considerations on the mode of deposition of the Rhaetic Series. *Quarterly Journal of the Geological Society, London*, **60**, 170–193.

SIEBOLD, C. T. E. von and STANNIUS, F. H. 1845. *Lehrbuch der vergleichende Anatomie*. Part 1. *Wirbellose Thiere, no. 1*. Berlin, 1–679.

SIMPSON, G. G. 1928. *A catalogue of the Mesozoic Mammalia in the Geological Department of the British Museum*. British Museum (Natural History), London, x+216 pp.

—— 1947. *Haramiya*, new name, replacing *Microcleptes* Simpson, 1928. *Journal of Paleontology*, **21**, 497.

SKELTON, P. W. and BENTON, M. J. 1993. Mollusca: Rostroconchia, Scaphopoda and Bivalvia. 237–264. *In* BENTON, M. J. (ed.). *The fossil record 2*. Chapman and Hall, London, xvii+895 pp.

SMITH, A. B. 1981. Implications of lantern morphology for the phylogeny of post-Palaeozoic echinoids. *Palaeontology*, **24**, 779–801.

—— 1990. Echinoid evolution from the Triassic to the Lower Liassic. *In* Rapport du Working-Group 'Limite Trias/Jurassique', Lyon – 28–29 Novembre 1988. *Les Cahiers de l'Université Catholique de Lyon, Série Sciences*, **3**, 79–117.

SMITH, W. E. 1957. Field Meeting in south Devon and Dorset. *Proceedings of the Geologists' Association*, **68**, 136–152.

SOLLAS, I. B. J. 1901. Fossils in the Oxford University Museum, V. On the structure and affinities of the Rhaetic plant *Naiadita*. *Quarterly Journal of the Geological Society, London*, **57**, 307–312.

SOWERBY, J. 1812–22. *The mineral conchology of Great Britain; or, coloured figures and descriptions of those remains of testaceous animals or shells, which have been preserved at various times, and depths in the earth*. Vol. 1, 1812–1815, 234 pp., +2 pp. index, +8 pp. supplementary index by Farey, J.; Vol. 2, 1815–18, 251 pp. (including the supplementary indices); Vol. 3, 1818–21, 194 pp.; Vol. 4, 1821–22, 116 pp. Meredith, London.

—— and SOWERBY, G. B. 1820–34. *The genera of Recent and fossil shells, for the use of students in conchology and geology*. 4 vols. Meredith, London. (Unnumbered. Commenced by James Sowerby ... and continued by G. B. Sowerby [1st of the name] ... with ... 267 plates by James Sowerby and J. de C. Sowerby).

SOWERBY, J. de C. 1824. *The mineral conchology of Great Britain*; *etc.* (continued). Vol. 5. Meredith, London, 444–485.

SPENGEL, J. W. 1881. Die Geruchsorgane und das Nervensystem der Mollusken. *Zeitschrift für wissenschaftliche Zoologie*, **35**, 33–383.

ST LAURENT, M. de 1980. Sur la classification et la phylogénie des crustacés décapodes brachyoures. *Comptes réndus hebdomadaires des Séances de l'Académie des sciences, Paris, Série D, Sciences Naturelles*, **290**, 1317–1320.

STEBBING, T. R. R. 1910. General catalogue of South African Crustacea. *Annals of the South African Museum*, **6**, 401–494.

STENSIÖ, E. A. 1916. Über einige Trias-Fische aus der Cava Trefontane, Tessin. *Bulletin of the Geological Institutions of the University of Upsala*, **15**, 13–34.

—— 1921. *Triassic fishes from Spitsbergen. Part 1*. Vienna, 307 pp.

—— 1932. Triassic fishes from East Greenland collected by the Danish Expeditions in 1929–1931. Part 1. *Meddelelser om Grønland*, **83**, 1–305.

STERNBERG, K. M. von 1833. *Versuch einer geognostisch botanischen Darstellung der Flora der Vorwelt (Skizzen zur vergleichenden Phytotomie vor-und jeztweltlicher Pflanzen Stämme, von A. C. I. Corda)*. Vol. 1, parts 5, 6. Fr. Fleischer, Leipzig, Prague, 80 pp.

STEVENSON, I. P. and MITCHELL, G. H. 1955. Geology of the country between Burton upon Trent, Rugeley and Uttoxeter. *Memoirs of the Geological Survey England and Wales*, 1–178.

STODDART, W. W. 1875. *Ceratodus forsteri*. *Proceedings of the Bristol Naturalists' Society, Series 2*, **1**, 145–149.

STOLICZKA, F. 1870–71. Cretaceous fauna of southern India. 3. The Pelecypoda, with a review of all known genera of this class, fossil and Recent. *Memoirs of the Geological Survey of India (Palaeontologica Indica), Series 6*, **3**, i–xxii +1–537.

STORRS, G. W. 1992. *Pachystropheus rhaeticus*, a choristoderan reptile from the

European Triassic and its implications for studies of the fossil record (Abstract). *Journal of Vertebrate Palaeontology*, **12**, 53A-54A.

—— 1993. Terrestrial components of the Rhaetian (uppermost Triassic) Westbury Formation of southwestern Britain. 447–451. *In* LUCAS, S. G. and MORALES, M. (eds). The nonmarine Triassic. *Bulletin of the New Mexico Museum of Natural History and Science*, **3**, 1–478+G1–G58.

—— 1994. Fossil vertebrate faunas of the British Rhaetian (latest Triassic). *Zoological Journal of the Linnean Society*, **112**, 217–259.

—— and GOWER, D. J. 1993. The earliest possible choristodere (Diapsida) and gaps in the fossil record of semi-aquatic reptiles. *Journal of the Geological Society, London*, **150**, 1103–1107.

—— —— and LARGE, N. F. 1996. The diapsid reptile *Pachystropheus rhaeticus*, a probable choristodere from the Rhaetian of Europe. *Palaeontology*, **39**, 323–349.

STRAHAN, A. and CANTRILL, T. C. 1904. Geology of the South Wales Coalfield, Part IV: the country around Bridgend. *Memoirs of the Geological Survey, England and Wales*, 1–120.

STRICKLAND, H. E. 1840. On the occurrence of a dragon-fly in the Lias of Warwickshire. *The Magazine of Natural History, London, New Series*, **4**, 301–303.

—— 1843. On certain impressions on the surface of the Lias bone-bed in Gloucestershire. *Proceedings of the Geological Society, London*, **4**, No. 92, 16–18.

—— 1845. *In* MURCHISON, R. I. *Outline of the geology of the neighbourhood of Cheltenham*. 2nd edition, revised by BUCKMAN, J. and STRICKLAND, H. E. John Murray, London, 68 pp., +Appendix of fossil remains, 65–109.

—— 1846. On two species of microscopic shells found in the Lias. *Quarterly Journal of the Geological Society, London*, **2**, 30–31.

STUTCHBURY, S. 1842. On a new genus of fossil bivalve shells. *Annals and Magazine of Natural History*, **1** (8, supplement), 481–485.

—— 1850. [On a large cylindrical bone found by Mr. Thompson in the 'bone-bed' of Aust Cliff, on the Severn.] *Report of the British Association for the Advancement of Science*, **1849**, 67.

SWEET, W. C. 1988. *The Conodonta: morphology, taxonomy, paleoecology, and evolutionary history of a long-extinct Phylum*. Oxford Monographs of Geology and Geophysics 10. Oxford University Press, Oxford, 212 pp.

SWIFT, A. 1989. First records of conodonts from the late Triassic of Britain. *Palaeontology*, **32**, 325–333.

—— 1995a. A review of the nature and outcrop of the 'White Lias' facies of the Langport Member (Penarth Group: Upper Triassic) in Britain. *Proceedings of the Geologists' Association*, **106**, 247–258.

—— 1995b. Conodonts from the Late Permian and Late Triassic of Britain. *Monograph of the Palaeontographical Society*, **149** (598), 1–80, pls 1–6.

SWINTON, W. E. 1948. Plesiosaurs in the City Museum, Bristol. *Proceedings of the Bristol Naturalists' Society*, **27**, 343–360.

SYKES, J. H. 1971. A new Dalatiid fish from the Rhaetic Bone Bed at Barnstone, Nottinghamshire. *Mercian Geologist*, **4**, 13–22.

—— 1974a. Teeth of *Dalatias barnstonensis* in the British Rhaetic. *Mercian Geologist*, **5**, 39–48.

—— 1974b. On elasmobranch dermal denticles from the Rhaetic Bone Bed at Barnstone, Nottinghamshire. *Mercian Geologist*, **5**, 49–64.

—— 1977. British Rhaetian bone beds. *Mercian Geologist*, **6**, 197–239.

—— 1979. *Lepidotes* sp., Rhaetian fish teeth from Barnstone, Nottinghamshire. *Mercian Geologist*, **7**, 85–91.

—— CARGILL, J. S. and FRYER, H. G. 1970. The stratigraphy and palaeontology of the Rhaetic Beds (Rhaetian: Upper Triassic) of Barnstone, Nottinghamshire. *Mercian Geologist*, **3**, 233–264.

SYLVESTER-BRADLEY, P. C. 1961. *In* BENSON, R. H. *et al.* Systematic descriptions. Q99–Q422. *In* MOORE, R. C. (ed.). *Treatise on invertebrate paleontology. Part Q. Arthropoda 3, Crustacea, Ostracoda.* Geological Society of America and University of Kansas Press, Boulder, Colorado, and Lawrence, Kansas, xxiii +442 pp.

SZANIAWSKI, H. 1974. Some Mesozoic scolecodonts congeneric with recent forms. *Acta Palaeontologica Polonica*, **19**, 179–199.

TAPPIN, D. R., CHADWICK, R. A., JACKSON, A. A., WINGFIELD, R. T. R. and SMITH, N. J. P. 1994. *United Kingdom offshore regional report: the geology of Cardigan Bay and the Bristol Channel*. HMSO, for the British Geological Survey, London, x+107 pp.

TASCH, P. 1969. Branchiopoda. R128–R191. *In* MOORE, R. C. (ed.). *Treatise on invertebrate paleontology. Part R. Arthropoda 4.* Geological Society of America and University of Kansas Press, Boulder, Colorado, and Lawrence, Kansas, xxxvi+398 pp.

TATE, R. 1864. The Liassic strata in the neighbourhood of Belfast. *Quarterly Journal of the Geological Society, London*, **20**, 103–111.

—— 1867a. On the Lower Lias of the north-east of Ireland. *Quarterly Journal of the Geological Society, London*, **23**, 297–305.

—— 1867b. Notes on the Liassic species of d'Orbigny in his 'Prodrome de Paleontologie'. *Geology and Natural History Repertory and Journal of Prehistoric Archaeology and Ethnology*, **1**, 394–396.

—— 1876. Class Lamellibranchia. 357–412. *In* TATE, R. and BLAKE, J. F. *The Yorkshire Lias.* J. van Voorst, London, 475 pp.

TAWNEY, E. B. 1866. On the western limits of the Rhaetic beds in South Wales and on the position of the 'Sutton Stone'. With a note on the corals by DUNCAN, P. M. *Quarterly Journal of the Geological Society, London*, **22**, 68–93.

TAYLOR, M. A. and CRUICKSHANK, A. R. I. 1993. A plesiosaur from the Linksfield erratic (Rhaetian, Upper Triassic) near Elgin, Morayshire. *Scottish Journal of Geology*, **29**, 191–196.

TERQUEM, O. 1855. Paléontologie de l'Étage inférieur de la Formation Liasique de la Province de Luxembourg, Grand-Duché (Hollande), et de Hettange, du Département de la Moselle. *Mémoires de la Société Géologique de France, 2ème Série*, **5**, 219–343.

—— 1866. *Cinquième mémoire sur les Foraminifères du Lias des Départements de la Moselle, de la Côte-d'Or, et de l'Indre.* Lorette, Éditeur-Libraire, Paris (Metz), 313–454.

—— and PIETTE, E. 1865. Le Lias Inférieur de l'Est de la France, comprenant la Meurthe, la Moselle, le Grand-Duché de Luxembourg, la Belgique et la Meuse. *Mémoires de la Société Géologique de France, 2ème Série*, **8**, 1–175.

THIELE, J. 1925. Gastropoden der deutschen Teifsee-Expedition II. *Wissenschaftliche Ergebenisse der Deutschen Tiefsee-Expedition 'Valdivia'*, **17**, 35–382.

THIOLLIÉRE, V. 1858. Note sur les poissons du Bugey, et sur l'application de la mèthode de Cuvier à leur classement. *Bulletin de la Société Géologique de France, 1ème Série*, **15**, 782–793.

THOMAS, J. B., MARSHALL, J., MANN, A. L., SUMMONS, R. E. and MAXWELL, J. R. 1993. Dinosteranes (4, 23, 24–methylsteranes) and other biological markers in dinoflagellate-rich marine sediments of Rhaetian age. *Organic Geochemistry*, **20**, 91–104.

TILLYARD, R. J. 1925. *The British Liassic dragonflies (Odonata).* British Museum (Natural History), 40 pp.

—— 1933. The panorpoid complex in the British Rhaetic and Lias. *In Fossil Insects No. 3.* British Museum (Natural History), 79 pp.

TINTORI, A. 1980. Teeth of the selachian genus *Pseudodalatias* (Sykes, 1971) from the Norian (Upper Triassic) of Lombardy. *Rivista Italiana di Paleontologia e Stratigrafia*, **86**, 19–30.

—— 1982. Hypsisomatic Semionotidae (Pisces, Actinopterygii) from the Upper Triassic of Lombardy (N. Italy). *Rivista Italiana di Paleontologia e Stratigrafia*, **88**, 417–442.

TOMES, R. F. 1878. On the stratigraphical position of the corals of the Lias of the midland and western counties of England and South Wales. *Quarterly Journal of the Geological Society, London*, **34**, 179–195.

—— 1884. A comparitive and critical revision of the Madreporia of the White Lias of the middle and western counties of England, and of those of the conglomerate at the base of the South Wales Lias. *Quarterly Journal of the Geological Society, London*, **40**, 353–375.

—— 1888. On *Heterastraea*, a new genus of Madreporia from the Lower Lias. *Geological Magazine*, dec. 3, **5**, 207–218.

—— 1903. Description of a species of *Heterastraea* from the Lower Rhaetic of Gloucestershire. *Quarterly Journal of the Geological Society, London*, **59**, 403–407.

TOMKIEFF, S. I. 1969. *Isle of Arran.* Geologists' Association Guide, 35 pp.

TORELL, O. M. 1870. Petrificata Suecana Formationis Cambricæ. *Lunds Universitets Årsskrift*, **6**, 1–14.

TRACEY, S., TODD, J. A. and ERWIN, D. H. 1993. Mollusca: Gastropoda. 131–167. *In* BENTON. M. J. (ed.). *The fossil record 2.* Chapman and Hall, London, xvii+895 pp.

TROEDSSON, G. 1951. On the Höganäs Series of Sweden (Rhaeto-Lias). *Lunds Universitets Årsskrift, Neue Folge, 2*, **47**, 1–269.

TRUEMAN, C. N. and BENTON, M. J. 1997. A geochemical method to trace the taphonomic history of reworked bones in sedimentary settings. *Geology*, **25**, 263–266.

TUTCHER, J. W. 1908. The strata exposed in constructing the Filton to Avonmouth Railway. *Proceedings of the Bristol Naturalists' Society, Series 4*, **2**, 5–21.

TUWENI, A. O. and TYSON, R. V. 1994. Organic facies variations in the Westbury Formation (Rhaetic, Bristol Channel, SW England). *Organic Geochemistry*, **21**, 1001–1014.

TYRRELL, G. Y. 1928. The geology of Arran. *Memoirs of the Geological Survey of Great Britain*, 1–292.

ULRICH, E. O. and BASSLER, R. S. 1908. New American Paleozoic Ostracoda. Preliminary revision of the Beyrichiidae, with descriptions of new genera. *Proceedings of the United States National Museum*, **35**, 277–340.

—— —— 1926. A classification of the toothlike fossils, conodonts, with descriptions of American Devonian and Mississippian species. *Proceedings of the United States National Museum*, **68**, 1–63.

USSHER, W. A. E. 1906. The geology of the country between Wellington and Chard. Explanation of Sheet 311. *Memoir of the Geological Survey, England and Wales*, i-vi+1–68.

VAN HEERDEN, J. 1979. The morphology and taxonomy of *Euskelosaurus* (Reptilia: Saurischia; Late Triassic) from South Africa. *Nasionale Museum, Navorsinge— Researches, Bloemfontein*, **4**, 21–84.

VAN WAGONER, J. C., MITCHUM, R. M., CAMPION, K. M. and RAHMANIAN, V. D. 1990. *Siliciclastic sequence stratigraphy in well logs, cores and outcrops: concepts for high resolution correlation of time and facies*. American Association of Petroleum Geologists Methods in Exploration Series, no. 7, 55 pp.

VAUGHAN, A. 1904. Palaeontological notes (a) Invertebrata (Mollusca). 201–208. *In* REYNOLDS, S. H. and VAUGHAN, A. The Rhaetic Beds of the South-Wales direct line. *Quarterly Journal of the Geological Society, London*, **60**, 194–213.

VOGT, C. C. 1852. *Bilder aus dem Thierleben*. Frankfurt-am-Main, vi+452 pp.

VOLZ, W. 1896. Die Korallenfauna der Trias monographisch bearbeitet. II. Die Korallen der Schichten von St.-Cassian in Süd-Tirol. *Palaeontographica*, **34**, 1–124.

VORABYEVA, E. I. and OBRUCHEV, D. V. 1964. Subclass Sarcopterygii. 268–322. *In* ORLOV, Y. A. (ed.). *Oznovii Paleontologii*, vol. 2. Navka, Moscow. [In Russian].

VYALOV, O. S. 1936. Sur la classification des huîtres. *Doklady Akademiya Nauk SSSR, Series 4*, **13**, 17–20.

WAAGEN, L. 1907. Die Lamellibranchiaten der Pachycardientuffe der Seiser Alm nebst vergleichend paläontologischen und phylogenetischen Studien. *Abhandlungen der Kaiserlich-Königlichen Geologischen Reichsanstalt, Wien*, **18**, i+1–180.

—— 1885. Salt Range fossils. Productus Limestone fossils, Brachiopoda. *Palaeontologica Indica, Series 13*, **1**, pt 4, 729–770.

WALTON, J. 1925. Carboniferous Bryophyta. I. Hepaticae. *Annals of Botany*, **39**, 563–572.

WANG GUANZHONG 1993. Xiphosurid trace fossils from the Westbury Formation (Rhaetian) of southwest Britain. *Palaeontology*, **36**, 111–122.

WARRINGTON, G. 1978. Appendix 1. Palynology of the Keuper, Westbury and Cotham beds and the White Lias of the Withycombe Farm Borehole. 22–28. *In* POOLE, E. G. Stratigraphy of the Withycombe Farm Borehole, near Banbury, Oxfordshire. *Bulletin of the Geological Survey of Great Britain*, **68**, 1–63.

—— 1981. The indigenous micropalaeontology of British Triassic shelf sea deposits. 61–70. *In* NEALE, J. W. and BRASIER, M. D. (eds). *Microfossils from recent and fossil shelf seas*. Ellis Horwood Ltd, for the British Micropalaeontological Society, 380 pp.

—— 1982. Palynology of cores from the basal Lias and the Permian(?)—Triassic sequence of the Winterbourne Kingston borehole, Dorset. *In* RHYS, G. H., LOTT, G. K. and CALVER, M. A. (eds). The Winterbourne Kingstone borehole, Dorset, England. *Report of the Institute of Geological Sciences of Great Britain*, **81/3**, 122–126.

—— and IVIMEY-COOK, H. C. 1990. Biostratigraphy of the late Triassic and early Jurassic: a review of type sections in southern Britain. *Cahiers Catholic University of Lyon, Science Series*, **3**, 207–213.

—— —— 1992. Triassic. 97–106. *In* COPE, J. C. W., INGHAM, J. K. and RAWSON, P. F. (eds). Atlas of palaeogeography and lithofacies. *Memoirs of the Geological Society, London*, **13**, i-ix+1–153.

—— —— 1995. The Late Triassic and Early Jurassic of coastal sections in west Somerset and South and Mid-Glamorgan. 9–30. *In* TAYLOR, P. D. (ed.). *Field geology of the British Jurassic*. Geological Society of London, Bath, ii+286 pp.

—— and WHITTAKER, A. 1984. The Blue Anchor Formation (Late Triassic) in Somerset. *Proceedings of the Ussher Society*, **6**, 100–107.

—— COPE, J. C. W. and IVIMEY-COOK, H. C. 1994. St Audrie's Bay, Somerset, England: a candidate Global Stratotype Section and Point for the base of the Jurassic System. *Geological Magazine*, **131**, 191–200.

—— AUDLEY-CHARLES, M. G., ELLIOTT, R. E., EVANS, W. B., IVIMEY-COOK, H. C., KENT, P. E., ROBINSON, P. L., SHOTTON, F. W. and TAYLOR, F. M. 1980. A correlation of Triassic rocks in the British Isles. *Special Report of the Geological Society, London*, **13**, 1–78.

WATERS, R. A. and LAWRENCE, D. J. D. 1987. Geology of the South Wales Coalfield, Part III, the country around Cardiff. 3rd Edition, Memoir for 1:50 000 geological sheet 263 (England and Wales). *Memoir of the British Geological Survey*, i–xi +1–114.

WATSON, H. 1930. On the anatomy and affinities of *Plicatula*. *Proceedings of the Malacological Society, London*, **19**, 25–30.

WEISHAMPEL, D. B. 1990. Dinosaur distribution. 63–139. *In*. WEISHAMPEL, D. B., DODSON P. and OSMÓLSKA, H. (eds). *The Dinosauria*. University of California Press, Berkeley, xvi+733 pp.

WELLS, J. W. 1956. Scleractinia. F328–F444. *In* MOORE, R. C. (ed.). *Treatise on invertebrate paleontology. Part F. Coelenterata*. Geological Society of America and University of Kansas Press, Boulder, Colorado, and Lawrence, Kansas, xx+F1–F498.

WENZ, W. 1938–1944. Gastropoda, Tiel 1. Allgemeiner Teil und Prosobranchia. *In* SCHINDEWOLF, O. H. (ed.). *Handbuch der Paläozoologie*, 6. Bornträger, Zelendorf, Berlin, 1639 pp.

WHALLEY, P. E. S. 1982. *Bintoniella brodiei* Handlirsch (Orthoptera) from the Lower Lias of the English Channel, with a review of British bintoniellid fossils. *Bulletin of the British Museum (Natural History), Geology Series*, **36**, 143–149.

—— 1983. A survey of recent and fossil cicadas (Insecta, Hemiptera—Homoptera) in Britain. *Bulletin of the British Museum (Natural History), Geology Series*, **37**, 139–147.

—— 1985. The systematics and palaeogeography of the Lower Jurassic insects of Dorset, England. *Bulletin of the British Museum (Natural History), Geology Series*, **39**, 107–189.

—— 1988. Mesozoic Neuroptera and Raphidioptera (Insecta) in Britain. *Bulletin of the British Museum (Natural History), Geology Series,* **44**, 45–63.

WHITESIDE, D. I. 1986. The head skeleton of the Rhaetian sphenodontid *Diphydontosaurus avonis* gen. et sp. nov. and the modernizing of a living fossil. *Philosophical Transactions of the Royal Society of London, Series B,* **312**, 379–430.

WHITTAKER, A. and GREEN, G. W. 1983. Geology of the country around Weston-super-Mare. Memoir for 1:50 000 geological sheet 279, New Series, with parts of sheets 263 and 295. *Memoir of the Geological Survey of Great Britain,* i–x +1–147.

WICKES, W. H. 1904. The Rhaetic Bone Beds. *Proceedings of the Bristol Naturalists' Society, Series 3,* **10**, 213–227.

WIGNALL, P. B. 1994. *Black shales.* Oxford University Press, 127 pp.

—— CLEMENTS, R. G. and SIMMS, M. J. 1989. The Triassic-Jurassic boundary beds of the City of Leicester. *Transactions of the Leicester Literary and Philosophical Society,* **83**, 25–31.

WILL, H. J. 1969. Untersuchungen zur Stratigraphie und Genese des Oberkeupers in Nordwestdeutschland. *Beihefte zum Geologischen Jahrbuch,* **54**, 1–240.

WILLEY, E. C. 1966. On the Mesozoic strata of the Cowbridge district, Glamorgan. Unpublished Ph.D. thesis, University College of South Wales and Monmouthshire, Cardiff.

WILLIAMS, M. E. 1972. The origin of 'spiral coprolites'. *University of Kansas Paleontological Contributions,* **59**, 1–19.

WILSON, E. 1882. The Rhaetic rocks of Nottinghamshire. *Quarterly Journal of the Geological Society, London,* **38**, 451–456.

WILSON, H. E. 1981. Permian and Mesozoic. *In* HOLLAND, C. H. (ed.). *A geology of Ireland.* Scottish Academic Press, Edinburgh, 335 pp.

—— and MANNING, P. I. 1978. Geology of the Causeway coast. 2 vols. *Memoirs of the Geological Survey of Northern Ireland,* i-vii+1–172.

WILTON, C. P. N. 1830. Memoir on the geology of the shore of the Severn, in the parish of Awre, Gloucestershire. *Quarterly Journal of Science, Literature and Art, London,* 64–73.

WINKLER, G. G. 1859. *Die Schichten mit Avicula contorta: inner- und ausserhalb der Alpen, etc.* Munich, 51 pp.

WITHERS, T. H. 1928. *Catalogue of the fossil Cirripedia in the Department of Geology. 1. Triassic and Jurassic.* British Museum (Natural History), London, xii+154 pp.

WOODWARD, A. S. 1888. A synopsis of the vertebrate fossils of the English Chalk. *Proceedings of the Geologists' Association,* **10**, 273–338.

—— 1889–1901. *Catalogue of the fossil fishes in the British Museum (Natural History).* 4 vols. British Museum (Natural History), London, 2493 pp.

—— 1889a. On some remains of fossil fishes from the Rhaetic beds of Spinney Hills, Leicestershire. *Transactions of the Leicester Literary and Philosophical Society,* **1**, 18–21.

—— 1889b. Note on a species of *Pholidophorus* from the Rhaetic Paper Shales of Wigston. *Transactions of the Leicester Literary and Philosophical Society,* **1**, 22–23.

—— and SHERBORN C. D. 1890. *A catalogue of British fossil Vertebrata*. Dulau, London, xxxv+396 pp.

WOODWARD, H. B. 1893. The Jurassic rocks of Britain, Vol. 3. The Lias of England and Wales (Yorkshire excepted). *Memoirs of the Geological Survey, England and Wales*, i–xii+1–399.

—— and USSHER, W. A. E. 1906. The geology of the country near Sidmouth and Lyme Regis. *Memoirs of the Geological Survey, England and Wales*, 1–96.

WRIGHT, A. D. and BENTON, M. J. 1987. Trace fossils from Rhaetian shoreface deposits of Staffordshire. *Palaeontology*, **30**, 407–428.

WRIGHT, T. 1860. On the zone of *Avicula contorta*, and the Lower Lias of the south of England. *Quarterly Journal of the Geological Society, London*, **16**, 374–411.

—— 1861. Monograph on the British Echinodermata from the Oolitic formations. Vol. 1, part 4. *Monograph of the Palaeontographical Society*, **12** (50), 391–468, pls 37–43.

—— 1874. Petrefacten aus der Raethischen Stufe bei Hildesheim, II. *Zeitschrift der Deutschen geologischen Gesellschaft*, **26**, 821.

—— 1880. Monograph on the British Echinodermata from the Oolitic formations. Vol. 2, part 3. *Monograph of the Palaeontographical Society*, **34** (158), i–iv, 155–203, pls 19–21.

ZENKER, J. C. 1836. *Historisch-topographisches Taschenbuch von Jena und seiner Umgebung, besonders in naturwissenschaftlicher und medicinischer Beziehung / Herausgegeben ... von J. C. Zenker*. Wackenhoder, Jena, x+338 pp.

ZEUNER, F. E. 1939. *Fossil Orthoptera Ensifera*. British Museum (Natural History), xiii+321 pp.

—— 1942. The Locustopsidae and the phylogeny of the Acridoidea (Orthoptera). *Proceedings of the Royal Entomological Society, London*, **B11**, 1–19.

ZIEGLER, J. 1964. Beschreibung einer Foraminiferenfauna aus dem Rhät vom Grossen Hassberg (Nordbayern). Bemerkungen zur Stratigraphie und Paläogeographie des Rhäts in Franken. *Geologica Bavarica*, **53**, 36–62.

ZIEGLER, P. A. 1982. *Geological atlas of western Europe*. Elsevier, Amsterdam, 130 pp.

ZIETEN, C. H. von 1830–33. *Die Versteinerungen Würtembergs, öder naturgetreue Abbildungen der in den vollständigsten Sammlungen, namentlich der in dem Kabinet des ... Dr. Hartmann befindlichen Petrefacten, mit Angabe der Gebirgs-Formationen, in welchen dieselben vorkommen, und der Fundorte*. Parts 9–12 (1833), 65–96. Stuttgart.

ZITTEL, K. A. von 1881–1885. *Handbuch der Paläontologie. I. Abtheilung*: *Paläozoologie. II Band. Mollusca und Arthropoda*. R. Oldenbourg, Munich and Leipzig, 893 pp.

SYSTEMATIC INDEX